Bits on Chips

Harry Veendrick

Bits on Chips

Second Edition

 Springer

Harry Veendrick
Heeze, The Netherlands

ISBN 978-3-030-09401-0   ISBN 978-3-319-76096-4   (eBook)
https://doi.org/10.1007/978-3-319-76096-4

Cover illustration: Wafer: Intel Corporation (USA)
                 Smart phone: HTC Corporation (Taiwan)

Printed on acid-free paper

This Springer imprint is published by the registered company Springer International Publishing AG part
of Springer Nature.
The registered company address is: Gewerbestrasse 11, 6330 Cham, Switzerland

Foreword

Microelectronics for everyone: that's the theme of this book. It discusses in a clear yet profound way the working principle of chips, their applications, and the way chips are made today. But why should people with little technical background be interested to know about all this? The answer is clear: *because "electronics for everyone" is also the theme of the twenty-first century.*

Chips are and will be used in everything surrounding us and enable us to make up a prosperous and sustainable world. A world where we cleverly consume renewable energy, where the modern diseases can be treated at a very early stage or even prevented, and where there are no boundaries in connecting with families and friends.

Chips in your household appliances will make them smart and make your life easier. Household appliances will decide by themselves when to use electricity, based on the current price and availability. Appliances at home communicate with the electricity grid which cleverly manages renewable energy sources such as solar and wind power.

Chips in your car will keep an eye on your speed and your level of concentration, and will help you to get to your destination as quickly and safely as possible. Your car will communicate with other cars to avoid accidents or to pass through information on the traffic or road situation. In the parking lot, or maybe even while driving, your car will exchange electricity with the smart grid.

Chips on the doctor's desk – a lab-on-chip – will allow the doctor to perform a fast and easy test for all kinds of diseases. With just a drop of blood, the doctor will be able to tell you whether you have a risk for cancer, a genetic disorder, etc.

Chips in your clothes will keep an eye on your heart problem, blood pressure, and brain waves (e.g., in case of epilepsy) with the aim to provide you with a thorough follow-up. Such wearable sensor systems will allow risk patients or older people to stay at home much longer than is the case today.

And everything will be connected to and through the *Internet of Things (IoT)*.

Let one thing be clear: *the key to all these wonderful applications will be the collaboration of scientists from various disciplines.* Engineers, physicists, chemists, biologists, sociologists, medical specialists, etc., will all add their specific point of view and professional capabilities to the problems we are facing today and together come up with solutions no one could have realized on their own.

So whether you are a future consumer using smart appliances or a scientist working together with engineers to make this smart environment into a reality, this book is an excellent starting point to unleash the full potential of the smart world of the future. Enjoy reading and enjoy the benefits of this technology in your everyday life.

Former CEO IMEC   Gilbert Declerck
Leuven, Belgium
March 2018

Preface

Integrated circuits (chips) are increasingly used to collect, process, store and transfer all types of information, make decisions, provide automation and control functions to deliver a significant contribution to the safety, health and wellness of mankind.

This book provides an insight into many aspects associated with integrated circuits, which are also generally referred to as micro-electronics. Topics include relevant design, implementation, fabrication and application aspects. Considerable effort has been invested to make the text both informative and accessible to a large audience. The large number of diagrams and photographs should reinforce the explanations. Some of the illustrations and photographs show conventional material, but then these are selected for educational purposes because of their clarity.

The idea to create this book was triggered by the request that I got more than a decade ago to present a one-and-a-half hour lecture on integrated circuits before a non-technical audience. This was quite a challenge: where should you start, where should you end and what should be in between? I have puzzled a long time with the same question for this book. If the text is too easy, it would not be of interest to people with a technical background. If the text is too technical, it would look like a course for graduates in semiconductor design, physics or fabrication. For this reason I have chosen to let this book be an introduction to both people with almost none to only little technical background and to people with a detailed technical background. Even the die-hards, the professionals with experience in one or more semiconductor disciplines, will find a lot of interesting material.

Therefore, this book contains two parts. Part 1 is specifically developed for people with almost no or little technical knowledge. It explains the evolution of (micro-) electronics, the basic components on a chip based on their applications and the similarity between a chip floor plan and a city plan. It continues with an explanation of frequently used semiconductor terminology, combined with a short discussion on the basic concepts of a transistor and the difference between analog and digital. It also explains the background behind the increasing complexity of electronic systems and their insatiable demand for processing power and memory storage. Finally a summary on the chip development cycle is presented. It discusses the basics of chip design, the chip's substrate (wafer), the lithography and fabrication process and the testing of the chips. The chip is one of the most complex technical devices that currently exist. It is therefore impossible to describe it without any technical term and certainly not within the limited number of text and illustrations pages of Part 1. Readers who need more information on certain subjects are referred to Part 2, in which most of the above subjects are discussed in more technical detail. Finally the book includes many pages of index, which enables the reader to quickly find the description of the various subjects, terminology and abbreviations used in the chip industry.

Not all information in this book is completely sprouted from my mind. A lot of books, papers and websites contributed in making the presented material state of the art. Considerable effort has been given to include all appropriate sources of and references to figures, photographs, diagrams and text. I apologize for possible faults or inaccuracies.

Acknowledgements

I greatly value my former professional environments: Philips Research labs and NXP Semiconductors. They offered me the opportunity to work with many internationally highly respected colleagues who were/are all real specialists in their field of semiconductor expertise. The experience built during my professional life, to which they also contributed, is one of the pillars below the creation of this book.

I very much appreciated the willingness of many companies and institutions which allowed me to use many colourful images to enrich this book with some of the most advanced devices, chips and systems.

I would like to thank my (former) colleague Joop van Lammeren for several nice photographs, his constructive discussions on the contents of this book and ideas for its title.

I would especially like to express my gratitude to my daughter Kim for the many hours she has spent on the creation of excellent and colourful artwork, which contributes a lot to the quality and clarity of this book.

I wish to thank my daughter Manon for her conscientious editing and typesetting work. Her efforts to ensure high quality should not go unnoticed by the reader.

Finally, my son Bram is thanked for his creative illustration of *Bits on Chips* on the cover of this book.

However, the most important appreciation and gratitude must go to my wife, again, for yet another 6 months of tolerance, patience and understanding, while creating this new edition.

Heeze, The Netherlands   Harry Veendrick
March 2018

Contents

About the Author

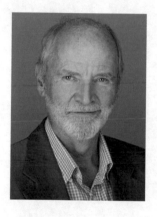**Harry Veendrick** joined Philips Research Laboratories in 1977, where he has been involved in the design of memories, gate arrays, and complex video-signal processors. His principle research interests include the design of low-power and high-speed complex digital ICs, with an emphasis on nanometer-scale physical effects and scaling aspects. Complementary to this is his interest in IC technology.

In 2002, he received the PhD degree in electronic engineering from the Technical University of Eindhoven, the Netherlands. He was a research fellow at Philips Research Labs and NXP Research and has been a Visiting Professor to the Department of Electronic and Electrical Engineering of the University of Strathclyde, Glasgow, Scotland, UK.

In 2006, he continued his research at NXP, which is the 2006 spin-off of the disentangled former Philips Semiconductors Product and R&D Departments. In May 2009, he left NXP Research and started his own training activity, teaching 1-day, 3-day, and 5-day courses for different target audiences (see: www.bitson-chips.com).

He (co-) authors many patents and publications on robust, high-performance, and low-power CMOS IC design and has contributed to many conferences and workshops, as reviewer, speaker, invited speaker, panellist, organizer, guest editor, and program committee member. In addition, he is the author of *MOS ICs* (VCH 1992), *Deep-Submicron CMOS ICs* (Kluwer Academic Publishers: 1st (1998) and 2nd edition 2000), and *Nanometer CMOS ICs*, 1st edition 2008 and 2nd edition 2017. He is a coauthor of *Low-Power Electronics Design* (CRC Press, 2004).

Part I
An Introduction to Microelectronics

Introduction

Part I is specifically developed for people with almost no or little technical knowledge. It presents an overview of the electronic evolution, the basics of chip development, some basic definitions and a variety of applications that use integrated circuits. Although a lot of attention has been given to clearly describe the topics, still some passages in this part may already sound a little too technical for the non-technicians, but then they need to accept that an integrated circuit is one of the most advanced technical products today. I hope that the reader is not discouraged by some of these passages but keeps on reading. Not everything needs to be fully understood to get a good impression of integrated circuits and build a good picture of their current and future potential in electronic systems.

Chapter 1 presents some history on the development of electronic components, starting with the vacuum tubes. This is followed by the invention of the first transistor and the first patent of the integrated circuit (IC, chip). Next, it also shows the evolution in growing complexity of the chips over the last six decades. It ends with a short summary on the impact of the chip on our current-day society.

Chapter 2 presents an overview of what a chip is. It discusses some typical application areas which require different grades of performance and reliability of their electronic components. These examples show why not everything can be done with a single chip. The chapter also discusses the similarity between a chip floor plan and a city plan, using metaphors to help explain concepts. It concludes a section on numbers that represent the exponential increase in performance, number of transistors per chip, etc.

Chapter 3 describes the basic concepts of the various circuits on a chip: the basic transistor, analog, digital, wireless, memories and interface circuits. These are combined with short definitions on the used terminology and necessary conversions: analog-to-digital conversion, decimal and binary numbers, bits and bytes and hardware and software.

Chapter 4 summarises the chip development cycle, starting with the design. After a short presentation on wafers, on which the chips are fabricated, it explains the basic steps of the photolithographic process to copy the detailed patterns of the design onto the various layers on the wafer from which the chips are built. Next, some of the process steps are discussed to create these layers. Finally when the fabrication process is completed, the individual chips on the wafer must be tested and the correct functioning ones must be encapsulated in a package. This chapter concludes Part I.

This does not mean that there is not more interesting stuff for readers with none or little technical background. By reading Part I, the reader will gain some insight into the various aspects of integrated circuits. This will enable him/her to also extract information from a large number of subjects that are discussed in Part II. For all categories of readers interested in bits and chips, I can only advise: scan through the subjects in Part II and find out what's of interest to you. A short summary on the contents of Part II can be found in the section Introduction to Part II. Also the exhaustive Index at the end of this book enables speedy access to desired topics, terminology and abbreviations.

For the readers that are less interested in more details about chip design, fabrication and test, I would advise to still read Chaps. 6, 14 and 15. Chapter 7 is important because memories form a major part of the total chip market and are essential parts in mobile phones, cameras, MP3 players, GPS systems, PCs, tablet computers, etc. Chapter 14 discusses the potentials of 3-D packaging of ICs, which has considerable impact on the size and form factor of electronic devices and systems. Chapter 15 presents the limitations of further scaling of integrated circuits and introduces the reader into the incredible potential of combining microelectronics (nanoelectronics) with nano- and biotechnology. This will result in the creation of a large range of new devices, sensors and actuators. Particularly the introduction of these devices in biotechnological and biomedical applications will create an enormous potential of compact electronic gadgets and systems that will impact human life, with respect to safety, environment, comfort, health and wellness far beyond today's imagination.

Chapter 1
The Electronics (R)evolution

1.1 Some History

An *integrated circuit* (*IC*), or more frequently called a *microchip* or *chip*, is a single piece of semiconductor material on which a large number of electronic components are interconnected. These interconnected components implement specific functions which are each tailored to a target application. The semiconductor material is usually silicon, but alternatives include germanium and gallium arsenide.

The chip is not just the result of an invention at some point in time. It is the result of a long evaluation process of electronic devices.

It all started with the invention of the *electron tube*, also called *vacuum tube* or valve, which consists of electrodes in a sealed glass enclosure in which the air is removed (*vacuum*). The most well-known electron tube was the three-electrode version, later called the *triode*, the original ideas of which were already patented in 1906 by Lee De Forest. The third electrode in this device was placed in between the cathode and the anode and consisted of a grid of small holes through which the electrons, emitted by the cathode, could pass and continue their path to the anode. This so-called *control grid* could manipulate the electron flow between cathode and anode and enabled an amplification of the control gate-gate signal onto the anode signal. This effect was first used in amplifying stages in radio transmitters and receivers. These tubes were very voluminous and required constant (indirect) heating of the cathode by a filament at the cost of large power consumption, even when used in microwatt amplification applications. Figure 1.1 shows a conventional TV receiver and its inside, built with electron tubes. Its power consumption was between 500 W and 1 kW.

The first electronic digital computer, the ENIAC, was assembled in 1945. It was a huge monster built from 30 separate units. With 18,000 vacuum tubes, hundreds of thousands of resistors, capacitors and inductors, and 1500 relays, it occupied about 1800 ft², weighed almost 50 tons and consumed about 200 kW. This produced a lot of heat, which was often the cause of malfunctions, although it used already some form of air-conditioning. Moreover, the vacuum tubes were fragile and

© Springer International Publishing AG, part of Springer Nature 2019
H. Veendrick, *Bits on Chips*, https://doi.org/10.1007/978-3-319-76096-4_1

Fig. 1.1 TV receiver (1955) and its inside using electron tubes. (Photo: Joop van Lammeren)

Fig. 1.2 Electron tubes, transistors (discretes) and chips

unreliable, which caused intolerably long downtimes of the computer. In comparison, its integrated circuit equivalent performing the same functionality, today, would operate very reliable on only 0.02mm^2 of silicon while consuming less than 1 μW.

With the invention of the *transistor* in 1947 by Bardeen, Brattain and Shockley (Bell Labs), the main problems with the vacuum tubes were solved. Initially, this solid-state device showed less performance but was much smaller in volume. Moreover, transistors turned out to be much cheaper and more reliable with a much longer lifetime. Although it took a couple of decades before the transistor showed a performance comparable to the vacuum tube, it soon replaced the vacuum tube for most low to medium performance applications. This was the drive for a rapidly increasing market of electronic products in the 1950s, with the transistor radio (1954) as the most well-known example. Figure 1.2 compares some electron tubes with transistors and chips.

A transistor radio was not only built from transistors; it also used resistors, capacitors and inductors. Initially, each of the *discrete devices* had to be placed,

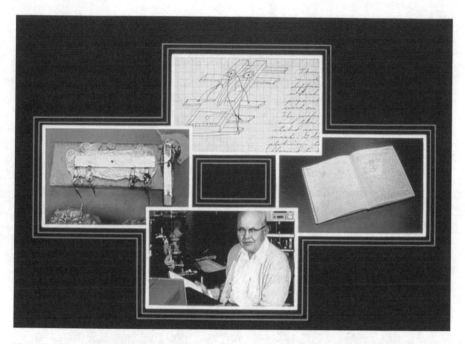

Fig. 1.3 The development of the first IC by Jack Kilby at Texas Instruments in 1958. (Photo: Texas Instruments/Koning & Hartman)

soldered and connected manually with wires on a board. Particularly for high-volume markets, this manual process was soon taken over by automatic 'pick-and-place' machines. But, with the growing complexity of these 'transistorised' products, also the application boards became larger in size. This caused increased length of the interconnections, requiring longer signal propagation time and increased power consumption. So, it was still impossible to build advanced computers with them.

A solution to this problem was found in 1958, when a researcher at TI, Jack Kilby, demonstrated the feasibility of combining resistors and capacitors with transistors on one single piece of semiconductor material (Fig. 1.3). He submitted a patent request entitled 'miniaturized electronic circuits' in 1959. Another researcher, Robert Noyce, at Fairchild Semiconductor, was working on similar things. Had Kilby focused his research on the making of the devices on a single substrate, Noyce's focus was to improve the connections between these devices. This resulted in the deposition and definition of metal tracks in an additional metal layer on top of these devices to connect them. These process steps became an integral part of the fabrication process. Both Kilby and Noyce are recognized for the combined invention of integrating many devices and connecting them effectively on a single substrate, which has become known as the '*integrated circuit (IC)*' or '*chip*'.

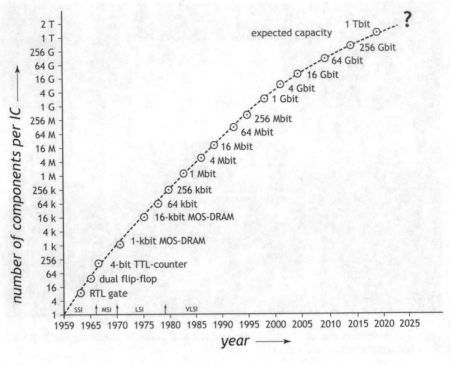

Fig. 1.4 Growth in the number of components per IC

1.2   Chip Evolution

So, an IC is one single piece of semiconductor material, on which a number of electronic components are connected together. These interconnected 'chip' components (usually transistors) implement a specific function. In most applications the semiconductor material is silicon, but alternatives include germanium and gallium arsenide. ICs are essential in most modern microelectronic products. Figure 1.4 illustrates the subsequent progress in IC complexity. It shows the number of components for advanced ICs and the year in which these ICs were first presented. Starting with the first chip in 1959, the line shows that this number is almost doubling every 2 years. Gordon Moore, who was also one of the founders of Intel, first recognized this trend in 1965. He predicted that this doubling would last for many generations of ICs to come. This prediction has become famous as *Moore's law*, which, although to a somewhat lesser extent, is still valid today, as shown by the complexity growth line, which is slowly saturating.

Integrated circuits come in various complexities and technologies. Discrete devices (*discretes*; see Fig. 1.2) are devices of very limited complexity, which are contained in their own package and do not share a common semiconductor substrate, as is the case with integrated circuits. To build a system with them, these individual devices need to be placed on a circuit board and connected together

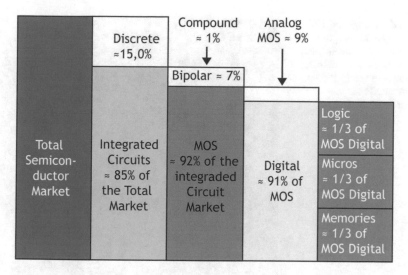

Fig. 1.5 Relative semiconductor revenue by IC category. (Source: IC Insights)

through the wires inside that board. Together with optoelectronic devices (e.g. LEDs and image sensors) and sensors, they still form about 19% of the total semiconductor market and include individual transistors, diodes, power transistors, RF/wireless products, basic logic gates, etc. Both discretes and ICs may be fabricated with different semiconductor technologies.

Figure 1.5 shows the relative semiconductor revenue per IC category. The most commonly used semiconductor technology is the *metal oxide semiconductor (MOS)* technology. The process of making an IC in it will be explained later. MOS ICs take about 80% of the total semiconductor market. Today's digital ICs may contain several hundreds of millions to more than a billion transistors on a single 1 cm² chip. They can be subdivided into three categories: general digital ICs (logic), microprocessors and memories. About 10% of the MOS ICs are of an analog nature.

Figures 1.6, 1.7, 1.8, 1.9, 1.10, 1.11, 1.12 and 1.13 illustrate the evolution in chip technology. Figure 1.6 shows a discrete (single) BC107 transistor (early 1960s), of which the majority of applications were found in audio amplifiers. This transistor is still used today!

One of the first ICs developed by Fairchild in 1961 is shown in Fig. 1.7. With only four transistors and two resistors, it is clearly the precursor of the modern integrated circuit. The transistors and resistors (light blue vertical bar at the right) are inside the silicon substrate, while the wiring is in an aluminium layer above the transistors.

The digital filter shown in Fig. 1.8 comprises a few thousand transistors. This chip (1979) was completely hand design and was used as one of the first digital filter applications in a TV set. It served to separate the luminance (brightness) from the chrominance (colour) information in a digital TV signal.

Fig. 1.6 A single BC107 bipolar general-purpose transistor for switching and audio low-power amplifier applications (early 1960s). (Photo: NXP Semiconductors)

Fig. 1.7 One of the first ICs developed in 1961. It is clearly the precursor of the modern integrated circuit. (Photo: Fairchild)

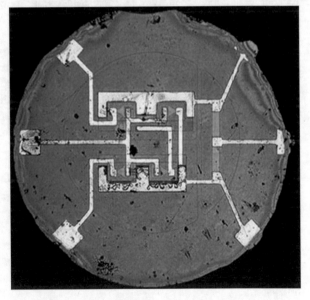

The digital audio broadcasting (DAB) chip (1997) in Fig. 1.9 performed digital audio decoding to enable a noise-free digital radio receiver and contained more than 4 million transistors. Today, DAB is a popular in-car technology that also includes real-time traffic and emergency information.

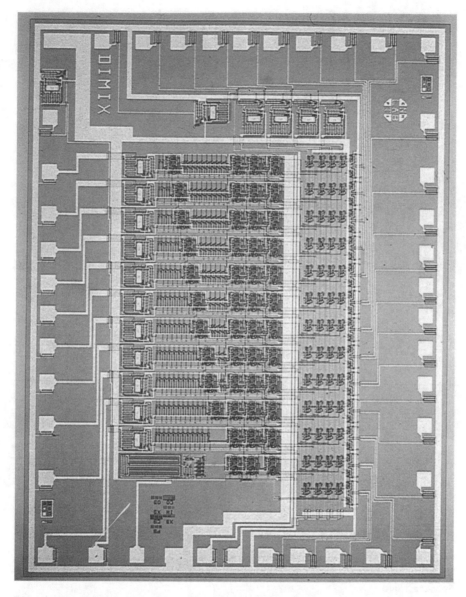

Fig. 1.8 A digital filter that separates the colour and brightness information of a video signal and comprises a few thousand transistors (1978). (Photo: NXP Semiconductors)

The Intel Xeon E5-2600 V3 18-core processor (2009) in Fig. 1.10 contains 5.69 billion transistors. The large areas with regular patterns represent the on-chip memories, which contain most of the chip's transistors. Next to these memories, this processor chip includes 18 identical computational engines (cores). The chip is targeted at Internet server applications.

Fig. 1.9 A digital audio broadcasting (DAB) chip, which contains some audio processing and various memories and comprises more than 4 million transistors (1997). (Photo: NXP Semiconductors)

Figure 1.11 shows the 16-core AMD Opteron 6380 series processor. The Zen-based Opteron will pack 32 cores on 2 of these dies mounted in a single package which will function as a single 32-core Opteron chip. With about 3.6 billion transistors, this chip has fewer transistors than the above cited Intel processor, mainly due to the much smaller on-chip memory capacity.

In the past, the continuous increase in microprocessor performance was primarily achieved by increasing the clock frequency, which almost approached 4 GHz. Above this level of frequency, the design complexity increased dramatically.

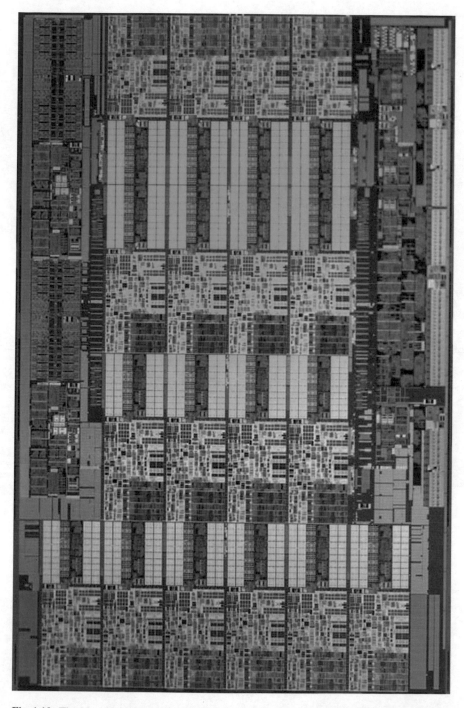

Fig. 1.10 The 14 nm-based Intel Xeon E5-2600 V3 18-core processor with 5.69 billion transistors on a 662mm² die, contains 45 MB memory and consumes 160 W for work station applications (2014). (Photo: Intel, Reprinted with permission)

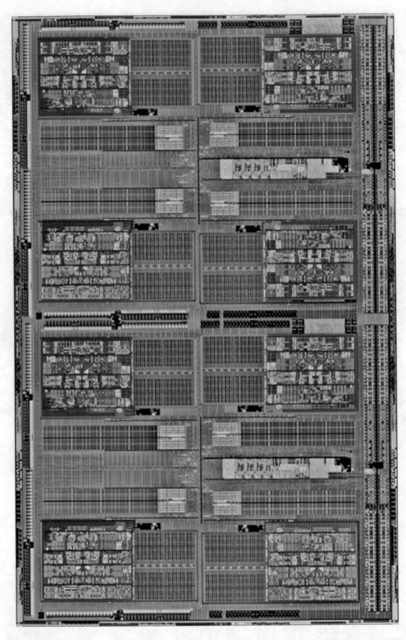

Fig. 1.11 The 14 nm FinFET-based AMD Opteron 6300 packs 8-core 'Piledriver' CPUs on a single die. The Zen-based Opteron will pack 32 cores on two dies mounted in one package which will function as a single 32-core Opteron chip (2016). (Photo: Advanced Micro Devices, Inc., Reprinted with permission)

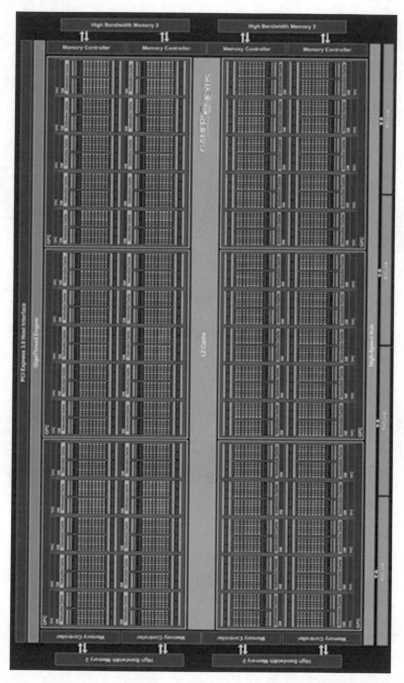

Fig. 1.12 The 16 nm-based Nvidia GP100 Pascal chip with 3584 CUDA cores containing 15.3 billion transistors on a 610 mm² large die, consuming 300 W for science-targeted accelerator cards. (© 2017 Nvidea Corporation, Reprinted with permission)

Fig. 1.13 A 256 Gb (more than 85 billion transistors), 3b/cell NAND flash memory chip in a 48-layer 15 nm CMOS technology. (© 2017 Sandisk, Reprinted with permission)

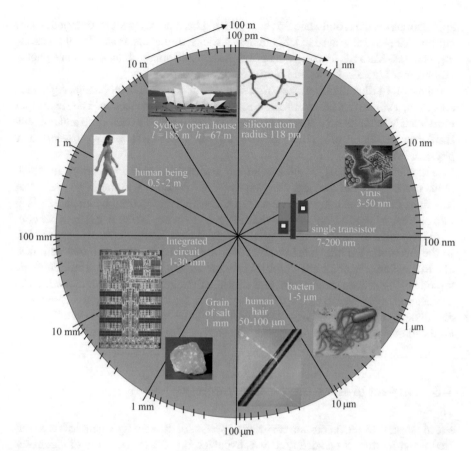

Fig. 1.14 Various semiconductor component sizes (e.g. atom, transistor, integrated circuit) in perspective

It became extremely difficult to master the timing behaviour of the hundreds of millions of signals across a chip and maintain proper signal propagation. Moreover, this had also led to unacceptable levels of power consumption; sometimes even more than 150 W for a single chip. Instead of increasing the clock frequency, microprocessor vendors are now increasing the number of computational engines, also referred to as *cores*, to achieve higher performance. The previous Intel and AMD processor chips are the result of this performance race in terms of multi-core processors.

The 16 nm GP100 Pascal chip of Nvidia with 3584 CUDA cores containing 15.3 billion transistors on a 610 mm^2 large die, consuming 300 W for science-targeted accelerator cards, is shown in Fig. 1.12.

Figure 1.13 shows a 256 Gb *non-volatile memory* chip containing more than 85 billion memory cells. The cells in this memory are so-called multi-level cells, which can store three bits each. A non-volatile memory chip keeps its information even when the supply voltage is switched off and is used in portable mass data storage applications such as solid-state drives, replacing hard drives in tablet computers and laptops. They are also used in USB memory sticks, digital cameras, global positioning systems, MP3 players, e-readers and mobile and smart phones, to store photographs, maps, music, books and phone directory, respectively.

Figure 1.14 illustrates the sizes of various semiconductor components, e.g. a silicon atom, a single transistor and an integrated circuit, in perspective. The sizes of an individual MOS transistor are already smaller than those of a bacterium and are now similar to the details of a virus, showing that the transistor sizes are close to their physical limit.

The above examples give already an impression of several applications in which chips can be used. In the next chapters we will come back to the huge variety of functions that can be integrated on a chip and which are dictated by an almost infinite number of applications. The increasing growth in number of transistors per chip is still expected to continue, although to a lesser extent, for one or two more decades. In the meantime the integrated circuits will be more and more combined with new electromechanical, chemical and/or biotechnical devices, sensors and actuators, which will introduce totally new features in existing consumer, automotive, communication, medical and military electronic equipment. The developments in nanoelectronics and nanophysics technologies will even lead to totally new applications.

1.3   Impact of the Chip on Our Society

From the start of the electronic revolution, some six decades ago, people have worried about its impact on society. It was expected that the introduction of electronic systems in industrial process automation would lead to massive labour reduction and layoffs. In the beginning, it certainly did. However, it was not anticipated, then, that the electronics revolution would even create more new jobs by developing totally new markets.

The backbone of this revolution is the chip. The rapid adoption of chip technology in industrial, military, aerospace, automotive, consumer, medical and communication markets created an electronics sector with the fastest growth rates in employment and trade of any industry over the last six decades. This has resulted in the fact that in the year 2017 the total turnover of the chip market, alone, exceeded 400 billion US$, while the worldwide electronics market is currently estimated to be close to 1700 billion US$ [1].

Even of more importance than just the amount of money involved is the impact of the chip on the society. It controls our everyday life in many aspects; e.g. it affects our working place, it improves safety and driving comfort of our cars, it enhances our listening (music) and viewing experiences (TV), and it makes many of our everyday tasks a lot easier and faster. But our social life is mostly affected by the rapidly growing information and communication markets. In this respect, the smartphone represents the summum of popular electronic products that combines entertainment with information and communication and enables to socialise with anyone, anytime and anywhere. These gadgets are marvels of electronic engineering, containing more features than an average PC, and their number will still increase. In combination with the Internet, these devices create the 'transparent world', where everybody has immediate access to numerous services and news sources worldwide. Local news has become global news in no time. Although chips are already used in many medical applications for several decades now, it looks like their adoption in medical and biological diagnostics and analysis has just begun. The creation of a lot of sensor devices, combined with the processing power of chips, will enable future applications in these domains that are even beyond our today's imagination.

All in all, chips improve safety, health and wellness of our everyday lives to such an extent that they have become the most dominant factor in the continuing electronic revolution. Part II, Chap. 15, presents more details on the future potential of microelectronics.

Further reading in this book will open the curtain to the fascinating world of integrated circuits (microelectronics; chips).

Other Interesting Stuff
The following websites present an overview of the used electronics terminology, application areas and timeline of inventions. This selection is subjective and not necessarily representative, but it is intended merely to supply the reader with a quick overview. In case these web sites are no longer accessible, a search on the Internet with the related entries will result in many interesting hits.

A lot of electronics terminology, historic dates and inventions can be found on: *www.electronicsandyou.com/electronics-history/history_of_electronics.html*

The importance of electronics with several application areas can be found on: *http://importanceofelectronics.blogspot.com/2009/06/history-of-electronics.html*

A timeline of technology inventions, many of which are related to electronics, can be found on: *http://www.explainthatstuff.com/timeline.html*

Chapter 2
What Is a Chip?

2.1 Chips Are All Around Us

An *integrated circuit* (IC) is basically an electronic circuit that integrates many different devices onto one single piece of silicon. It is also often called a *microchip* or simply *chip*. Generally the word chip refers to a small and very thin piece of material, sometimes broken from a larger piece of material. When a chip is not yet packaged, we call it a *die*. So, most integrated circuits are produced as dies on very thin (less than 1 mm thick) silicon wafers (Fig. 2.1). Usually 25 or 50 wafers are produced in one *batch* or *lot*. Specialists would immediately recognise the conventional 6 inch wafer, but the figure perfectly illustrates the different phases of a chip.

The small rectangular chips on the wafer are surrounded by so-called narrow *scribe lanes*. These are used to separate the chips from each other by sawing or laser cutting, after they have been fabricated and fully tested. Next the correct chips are put in an encapsulation, which is usually called *package*. Then they are tested again for correct functioning. The correct chips are then mounted onto a *printed circuit board*, e.g. a small printed circuit board in a mobile phone, or a motherboard in a PC, laptop or tablet PC.

Microchips, which are often less than 5 mm on a side, are all around us. The impact of these tiny chips on our daily lives is huge. They have a controlling function in applications as traffic lights and industrial process control. They add convenience by smoothly regulating our room temperature but also by enabling cruise control and GPS in our cars. They support health by easing medical diagnostics and monitoring processes. Finally they increase our feeling of safety by enabling night vision and smoke detection in and around our homes as well as tyre pressure monitoring and anti-lock braking systems (ABS) when driving our cars. In general they improve quality of life even when we are unaware of their presence. In the morning we wake up with a song by Michael Jackson, and then we take a shower in our preheated bathroom, heat some food in the microwave, listen to the news, look at our watch, make a phone call, use the electronic car key, start the car, program the GPS, set the cruise control, select the broadcasting station or play MP3 and drive off.

H. Veendrick, *Bits on Chips*, https://doi.org/10.1007/978-3-319-76096-4_2

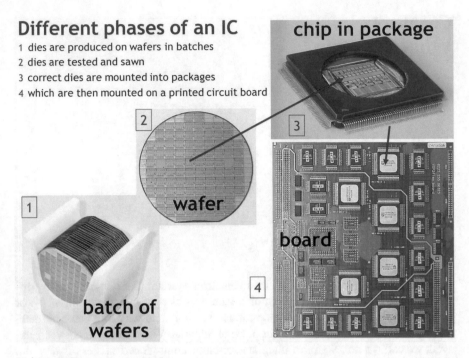

Fig. 2.1 Different phases of a chip

After parking the car using the rear parking sensors, we take our employee RF ID card to get access to the building, take the elevator, enter the office and start the computer, after one or more phone calls. By the morning coffee break, most of us have already come into contact with 50 or more of these microchips.

So, not only has the chip become an essential product in all existing electronic products, it also enables product improvements and the creation of totally new products at an incredible rate.

This all clarifies that the invention of the chip has been identified as one of the most important innovations of mankind, if not *the* most important.

2.2   Why Cannot Everything Be Done with a Single Chip?

As discussed before, the number of chip applications is huge and still expanding. The major reason that not all of these applications can be served by one and the same chip is that there are stringent requirements that are characteristic for an application domain. ICs designated for passive RFID cards, for instance, have no power supply. In such applications, the power is induced by an external source (card reader) that also initiates signal transmission when moving the card close to the reader. This means that the ICs on these cards must be extremely low power. Such

applications also require a high level of security and privacy, particularly in banking
and electronic passport applications.

ICs for automotive applications, on the other hand, require a high level of robustness because of the hostile environment (e.g. dirt, humidity, vibration, large temperature variations and large noise levels caused by the engine ignition) with high
reliability standards (e.g. for electronically controlled brake systems, tyre pressure
monitoring and airbag control).

On its turn, the mobile application domain has quite different characteristics and
requirements, which will be extensively discussed in one of the next paragraphs. It
is clear that all these different application characteristics require a variety of electronic circuits and solutions, customised to the needs of the application.

2.3   What's on a Chip?

A chip, or integrated circuit (IC), is a composition of many basic functions (*building
blocks*) and interfaces to the outside world (*I/Os*). The way these building blocks are
interconnected and how they communicate with each other determines the behaviour and performance of the chip. The functions that are needed on a chip depend on
the requirements of the application for which it is intended. A well–known electronic application is that of a digital thermostat (Fig. 2.2) for controlling the temperature in a living room, for example.

In fact, this thermostat contains a temperature sensor, a controller chip and a
display. The user can pre-set various temperatures for different time periods over the
daytime (morning, evening, night, etc.), over the weekdays (Monday, Tuesday, ...,

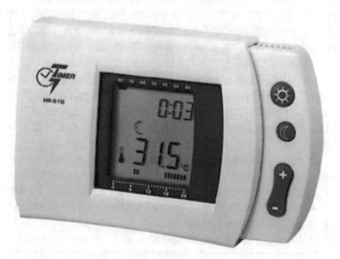

Fig. 2.2 A well-known application showing its specific functionality. (Photo: Advanced Timer
Technologies Ltd)

weekend), for different seasons (summer, winter, holidays), etc. The sensor measures the room temperature and feeds it to the chip, on which it is compared with the pre-set temperature that the user has stored into the on-chip memory. This is continuously done during the whole day. Because the temperature sensor delivers an analog signal, it first needs to be digitised by an analog-to-digital (A/D) converter, before it can be compared with the digitally stored pre-set temperature. So, basically the chip contains an A/D converter, a memory to store the temperature pre-settings and some of the pre-installed summer, winter and holiday mode settings, an electronic clock, a small processor that compares the actual temperature with the pre-setting and checks the battery state and a few drivers that need to drive the LCD display.

The above pre-setting and control functions enable a fully automated temperature regulation 24 h a day, 365 days a year. These relatively simple functions can easily be implemented on a very small and cheap chip.

2.4   From Required Functionality to Required Operations

Another well-known example of a complex electronic system is the mobile 'phone', which has evolved to a multifunctional gadget which is currently often referred to as *smartphone* (Fig. 2.3). Today, its functionality, which is also often referred to as number of *features*, is so large that we can hardly call it a phone anymore. In fact, it is a digital gadget that can do many of the things that were previously done by dedicated single-function electronic gadgets.

Table 2.1 shows an incomplete list of 'smartphone' features and the basic operations that are required for executing the respective individual feature.

Let us take the example of the digital camera function. First, the picture is captured by the image sensor. Next it is converted from the analog to the digital domain and then stored in the cache memory as a file. Then this file is copied (stored) into the removable memory card (e.g. a secure digital (SD) card, compact flash card (CF), etc.). The colour LCD screen allows review of the captured images. In that case, they first have to be read from the memory and then displayed on the colour screen. After completion of a photo session, the memory card can be taken out of the camera and inserted into the card reader of a PC, on which a backup of these pictures can be stored.

To realise the various features with a smartphone, it combines various electronic circuits with physical components. Advanced smartphones include two cameras, one on the front that may serve as a webcam and one on the back for taking pictures and videos. Several smartphones are equipped with a gyroscope that enables motion sensing, user acceleration and rotation.

A receiver/transmitter is needed to deal with SMS messages and phone calls and to perform Bluetooth, Internet and satellite (GPS) communication. Next it needs analog-to-digital conversion and vice versa e.g. to convert voice to digital format and back. Digital signal processing is needed, for instance, for electronic camera

Fig. 2.3 Example of the functionality of an advanced smartphone. (Photo: HTC)

Table 2.1 Modern mobile phone functions (features) and their required operations

Function (feature)	Operation
Phone	Receive, amplify, speaker, send
SMS	Receive, type, send
Answering machine	Receive, store, read, display, send
Personal organizer and calculator	Type, store, read, display
Connectivity: e-mail, internet, GPS, blue-tooth	Receive, type, send, display
GPS	Receive, read, display
Music (MP3)	Store, read, decode, play
Digital camera + camcorder	Capture, store, read, display
WEBcam	Capture, send
Ring tones	Download, store, read, play
Games	Store, read, execute, display
Radio	Receive, amplify, play
Wireless modem for PC	Receive, send, display
Clock and wake-up calls	Timer, display, play

zoom and MP3 decoding. A control processor serves as a display interface. However, it is also used to check the battery state, to perform the phone menu control and to monitor the key pad pressure. A *SIM card* is a standard memory card used in mobile phones to store personal identity information, cell phone number, phone book, text messages and other personal information. It is a portable memory chip that automatically initialises the phone into which it is inserted. Other memory chips are used to store SMS messages, a GPS map + route, MP3 music, pictures and video, ring tones, voice recorder, etc. Several power (management) modules are needed to regulate and control the power supply of the various electronic components.

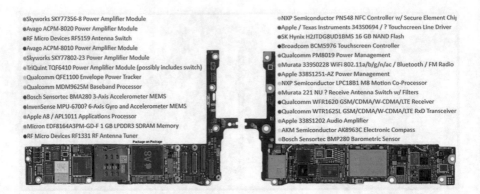

Skyworks SKY77356-8 Power Amplifier Module
Avago ACPM-8020 Power Amplifier Module
RF Micro Devices RF5159 Antenna Switch
Avago ACPM-8010 Power Amplifier Module
Skyworks SKY77802-23 Power Amplifier Module
TriQuint TQF6410 Power Amplifier Module (possibly includes switch)
Qualcomm QFE1100 Envelope Power Tracker
Qualcomm MDM9625M Baseband Processor
Bosch Sensortec BMA280 3-Axis Accelerometer MEMS
InvenSense MPU-6700? 6-Axis Gyro and Accelerometer MEMS
Apple A8 / APL1011 Applications Processor
Micron EDF8164A3PM-GD-F 1 GB LPDDR3 SDRAM Memory
RF Micro Devices RF1331 RF Antenna Tuner

NXP Semiconductor PN548 NFC Controller w/ Secure Element Chip
Apple / Texas Instruments 343S0694 / ? Touchscreen Line Driver
SK Hynix H2JTDG8UD1BMS 16 GB NAND Flash
Broadcom BCM5976 Touchscreen Controller
Qualcomm PM8019 Power Management
Murata 339S0228 WiFi 802.11a/b/g/n/ac / Bluetooth / FM Radio
Apple 338S1251-AZ Power Management
NXP Semiconductor LPC18B1 M8 Motion Co-Processor
Murata 221 NU ? Receive Antenna Switch w/ Filters
Qualcomm WFR1620 GSM/CDMA/W-CDMA/LTE Receiver
Qualcomm WTR1625L GSM/CDMA/W-CDMA/LTE RxD Transceiver
Apple 338S1202 Audio Amplifier
AKM Semiconductor AK8963C Electronic Compass
Bosch Sensortec BMP280 Barometric Sensor

Fig. 2.4 Top main board of the Apple iPhone 3G. (Source: Apple Inc; components are labelled and noted by Semiconductor Insights)

Finally, interface devices such as keyboard, microphone, stereo headset and colour display complete the list of components.

The different features may be performed by a combination of ICs from various vendors. Figure 2.4 shows the front side and the back side of the iPhone6 board as an example. It also gives an indication of the overall system complexity of a smartphone today. It consists of many different components from many different vendors, and they all need to communicate with each other in a reliable way. In addition, some of these components are systems in itself. The Apple iPhone X generation, introduced in 2018, includes an A11 chip with six cores and 4.3 billion transistors. Its processing power is comparable or even more than that of an Apple iMAC computer or MacBook laptop.

The accelerated increase, over the last decade, in the number of features and apps that can be performed by a smartphone was mainly due to rapid advances in chip process technology and performance.

Another fast-expanding chip application domain is the automotive electronics in and around the car: engine management, safety, in-car infotainment, driver assistance, passenger comfort and autonomous driving. Typical cars, today (2018), therefore contain between 600 and 1000 chips. More details on automotive systems are presented in the final chapter in this book, as they rely on a large number of sensors

Although the functions on a thermostat chip, a camera chip, mobile phone chip or automotive chip are very specific to the application, we can create a kind of generic chip architecture for educational purposes. In the next section this will be discussed in more detail.

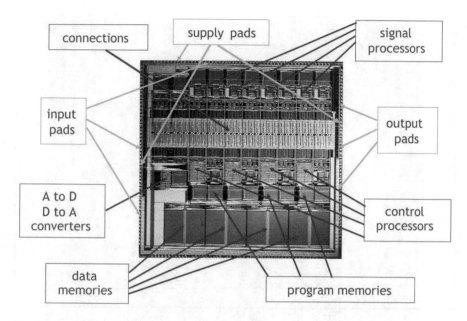

Fig. 2.5 Example of basic building blocks, interconnections and interfaces on a chip

2.5 Basic Building Blocks on a Chip

In this section we will only position several building blocks on a chip. A discussion of the functionality of these blocks on a chip is moved to Chap. 3, together with a description of the basic MOS transistor, from which 90% of all integrated circuits are built.

In a generic representation (Fig. 2.5), a chip may consist of a variety of functional *building blocks*: analog-to-digital (ADC) and digital-to-analog (DAC) converters; control processors, also called *microcontrollers* or *microprocessors*; and digital signal processors, which perform the processing of, e.g. digital audio or video signals. Next, the results (data) from these building blocks need to be stored into one or more memories. In certain applications the program that must be executed by the chip is stored in a so-called program memory. All building blocks and memories communicate through many interconnections. Finally, the chip needs to be supplied through supply pads, and it needs to communicate with other parts of the system such as other chips, a keyboard or a display. This is performed through the chip interfaces, which are often called I/O (input and output) pads.

The floor plan and layout of the chip are made such that all components together occupy the smallest silicon area. The design of an integrated circuit is a very conscious, precise and time-consuming task. To create a feeling for the chip complexity, we use the development plan of a city as a metaphor to illustrate the development of a chip, because of their many similarities.

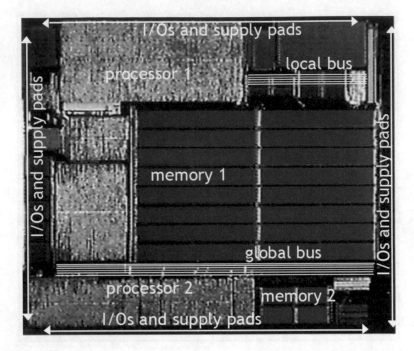

Fig. 2.6 Basic example of a chip floor plan

2.6   Similarity Between Chip Floor Plan and City Plan

Suppose a new 1-million-inhabitants city, including its urban, shopping and industrial areas, its complete infrastructure (electricity, gas and water, roads, highways, railways and airports) needs to be completely developed from scratch.

First it must be decided where to locate the urban area. Then the industrial area must be developed at some distance from the urban area. Also the locations of distribution centres, shopping centres, train, bus and subway stations, roads and motorways and airport are very important in order to guarantee an efficient access to these public facilities. Next the long-term supply requirements are an essential part of the total city infrastructure.

There are many similarities between the development of a city plan and a floor plan of a chip. Figure 2.5 showed an example of various components (building blocks, connections and pads) on an average chip, each with its own specific functionality. Each building block (*core*) is individually designed or available from a library. After the design of all building blocks is completed, a floor plan is created to fit them together with the connections and interface pads on the smallest silicon area.

In the following we will successively discuss the similarities between a city plan and a chip floor plan in more detail on the basis of Fig. 2.6.

A chip can contain a large number of building blocks, sometimes more than 50. In this example we limit the number of building blocks to only a few different memories and different processors, to keep the discussion simple. The following list explains these similarities for each of the blocks of the chip in Fig. 2.6.

- *A production plant is similar to a certain kind of processor: processor1*

   Let us take a production plant (fab) for aluminium cans (for beverages) as an example. This fab receives aluminium material in the form of sheets and *processes* it into cans. Next, these cans are tested to check for potential leaks and then cleaned, coloured and printed. We can compare all these processes to produce cans with processes (e.g. add, multiply, code, etc.) performed by an on-chip processor, which we will call processor1.

   Then the cans are packaged in paper sleeves and palletised for shipment to a huge storehouse.

- *A storehouse or distribution centre is similar to a certain kind of memory: memory1*

   The cans in the storehouse *are stored* for a certain period of time, before they are transported to a filling factory. The amount of storage time is determined by when the filling factory is running out of cans. This storage time could be as long as weeks or months. We may compare the function of a storehouse with that of an on-chip memory, which we will call memory1. The data in this memory (e.g. the result of the add, multiply or code operation) is stored until it is needed by another building block (processor2).

- *A filling factory is similar to another kind of processor: processor2*

   When the cans arrive from the storehouse into the filling factory, they are de-packaged, extensively cleaned and send to the filling machines. The filled cans are mechanically sealed, coded with the filling date and printed with the 'best before' date. After arranging them into a multi-pack format, they are shrink-wrapped and again palletised. The total filling process requires various operations, which are quite different from the can production process. Back to the chip analogy, this means that we need a second processor, processor2, because its functionality (e.g. compare, filter, decode, etc.) is quite different from processor1.

   Then the cans are distributed to stores and supermarkets.

- *A store or supermarket is similar to another kind of memory: memory2*

   Usually the storage time of most products in a store or supermarket is limited by the 'best before' date to only a few days. This storage function can be compared with that of another on-chip memory: memory2. This one certainly has a different access and storage time and sequence compared to memory1, because consumers can buy these products (cans) every day, every hour or every minute.

- *The supply (electricity, water, gas, etc.) network is similar to the on-chip supply network*

All production plants, factories, distribution centres, stores and supermarkets must be connected to electricity, gas and water, which are very similar to the on-chip supply network. The city needs a well-developed supply infrastructure to meet the surge in power demand during peak hours. Similarly, the chip requires a very robust supply network to be able to supply the extreme current surges, without too much voltage loss during peak activity.

- *The roads, highways and freeways are similar to the on-chip communication lines: local and global buses*

A good infrastructure of freeways, highways, main roads and small roads must enable traffic accessibility to all parts of the city at sufficient speed. To support the increasing traffic capacity, the highways will consist of many parallel lanes, sometimes even 12 on each side. This is very well comparable to the on-chip interconnections. Usually the communication on a chip is done through many parallel signal lines. This is called a *bus*. Today we see 16-bit, 32-bit or even 64-bit buses on a chip. They may support only local connections, *local bus*, or long-distance interconnections, *global bus*.

- *At the edges of the city, these networks are connected to the global supply infrastructure with terminals: supply terminals*

Electric power, for example, reaches our city via high voltage transmission lines to reduce the amount of power loss (Joule heating) in the global network. With power transformers, these voltages are reduced, e.g. from 10,000 V to 110 V or 220 V. The lower voltage is then only locally distributed in the city. This is similar to the supply terminals on a chip (*supply pads*), which transfer the voltage to the inner circuits.

- *Long-distance transportation between cities uses interstate highways (freeways), harbours, train stations and airports, similar to the chip input and output terminals: I/O pads*

The choice, which means of transport is taken, depends on the distance that has to be travelled, on what has to be transported and on how quickly it has to be at the other location. The same holds for the I/O terminals on a chip. There may be chip interfaces running at an average speed, comparable to taking the train. High-speed communication, like a DDR memory interface (see Part II, Sect. 6.5), is similar to taking an airplane.

- *Finally, traffic lights hold the traffic on red and let it go on green, similar to a clock signal on a chip.*

Although traffic lights do not contribute to the amount of transport of persons and material, it certainly contributes to a better organised and safer way of traffic. On green, the traffic starts to travel to the next traffic light. This is analogous to a clock signal on a chip that controls the flow of information (bits) across a chip. This is done by routing the clock signal to flip-flops, which operate similar to the traffic lights in a city. When the clock signal turns high, the *flip-flop* output signal travels through the logic gates and should arrive at the next flip-flop within one clock period.

Now we have explained the similarities between a city plan and a chip floor plan, we will extend the analogy even further and compare the development of the various building blocks on a chip with the building process of homes or commercial properties in the city.

2.6.1   Similarity Between Building a House and Creating a Chip

When building a home or store, individual bricks to construct its walls (Fig. 2.7) could be used. One can see that this wall is a purely manual construction, taking a lot of labour and time, particularly when the home is big.

Because both labour and time are very costly today, the building process is much more automated. Instead of using separate bricks to build a wall, complete

Fig. 2.7 Individual bricks used in the construction of a wall

Fig. 2.8 Improved building efficiency by using prefab walls

prefabricated walls are used to increase the efficiency of the building process. Figure 2.8 shows the placement of such a prefab walls.

To further increase the efficiency of the building process, completely prefabricated homes (Fig. 2.9) are available, which reduce the 'building time' to only the placement of the various building blocks, like living room, kitchen and bedrooms and connect them to the supply infrastructure of the city. However, the higher the level of prefabrication, the less flexibility in deviations from the standard concept of the home is allowed.

As mentioned before, building a home is very similar to the design of an on-chip building block. We can create that building block by using many individual transistors as shown in Fig. 2.10a.

Today's ICs' complexity has surpassed the billion transistor level. If we would have to design every building block on a chip from scratch, with individual transistors, analogous to building a wall from individual bricks (Fig. 2.7), it would take an incredible amount of design time. So, again similar to building a home, we need to improve the designer's efficiency. This can be done by creating the building block either from predesigned *standard cells* (Fig. 2.10b), available from a library, analogous to the prefab wall in Fig. 2.8, or using a *predesigned building block* also available from the library (Fig. 2.10c), analogous to the pre-fabricated living room, kitchen or bedroom to construct the house in Fig. 2.9. We also call such a prefabricated functional block an *IP building block* (IP = intellectual property), which expresses that it contains a lot of intellectual property rights and may only be used

Fig. 2.9 Further improved building efficiency by using completely prefabricated homes. (wee-House®, 2018)

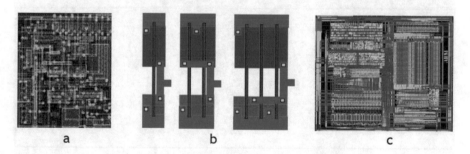

Fig. 2.10 (**a**) Individual transistors (similar to the bricks in Fig. 2.7) and (**b**) predesigned standard cells (similar to the prefab walls in Fig. 2.8) and (**c**) a predesigned building block (similar to the prefab home in Fig. 2.9)

under a licence agreement, if it is not an in-house developed building block. Predesigned building blocks can only be applied if its functionality matches the application requirements.

As explained before, we can extend the similarities even further. Even the roads and highways have their electronic counterparts on the chip (Fig. 2.11). These are similar to the on-chip buses, in which one bus line (metal wire) on a chip is comparable with one lane of a highway. The trend towards more parallel lanes and more multi-level crossings in highways is also visible on the chip. The lower part of the figure shows several metal line crossings on a chip in five different metal levels (layers).

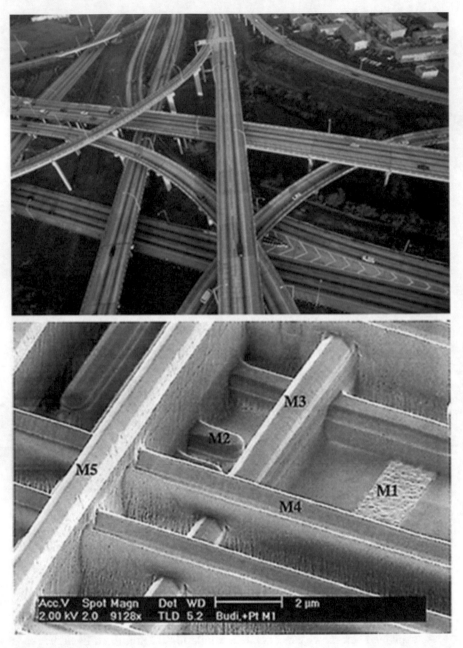

Fig. 2.11 Similarity between highways and crossings (above) and on chip bus lines and crossings (below)

Today's highways may have 10 to 16 lanes, while the buses on a chip may even be 32 or 64 bits wide. While five to six levels of highway crossings can be found today close to or in big cities, on a chip the number of layers has already increased to more than ten. Also the traffic speed on the highways is similar to the speed of the signals over the on-chip metal wires.

In conclusion, there are a lot of similarities between the design of an IC and the development of a city. The big difference between these two is that the total development time of a full chip is only 1 or 2 years.

2.6.2   However…

Over the last four decades, the average sizes of all components on a chip scaled with a factor of 0.7 about every 2 years. As a result, the number of transistors on a same size chip has doubled every 2 years. This is known as *Moore's law*.

Returning to our city development, this would mean that *every 2 years*:

- We have to double the number of homes and fabs, without increasing the city area.
- We have to increase the fab production output by 40%.
- We have to speed up the traffic throughput by 40%.
- And use about the same energy for all of this.

To translate this trend in real numbers and only consider the last three decades, then *in this 30-year period*:

- The number of transistors on a chip has increased by a factor of 10,000 and their speed by a 100 times.
- The chip functionality has therefore increased by a factor of 1 million.
- A highway lane would have shrunk 60 times in width with the same or higher traffic throughput.

Moreover, on an advanced processor chip for PC or laptop, today (2018), the number of transistors is approaching the 10 billion mark. The number of transistors on a smart phone chip is rapidly increasing as well and is approaching the 5 billion mark. All of these transistors need to be interconnected by wires. The total length of these wires approaches 2 miles (>3 km).

These numbers show the incredible progress that the semiconductor industry has made over the last four to five decades. It also enabled the electronic gadgets to increase their functionality so quickly over the last two decades. This is also related to the lifetime of a good or product. A home looks obsolete after 30–50 years, a car after 10 years, a TV set after 5 years but a mobile phone after only 2 years.

2.7   Conclusion

This chapter presented a flavour of the large variety of functions and features that can be integrated on a chip. It also introduced several examples of application domains that have their own specific requirements with respect to the applied integrated circuits. While costs are a major driver to keep consumer chips small, battery lifetime is the major driver to keep power consumption of mobile chips low, and safety is the major driver for the high reliability standards for automotive, medical, military and aviation chips.

Now that we have a reasonable insight into what a chip actually is, we will spend some more time, in the next chapter, on the basic circuit concepts of the various functions on a chip.

Chapter 3
Basic Concepts and Definitions

3.1 Introduction

The world of integrated circuits is a very diverse and complex one. Many disciplines are involved in their development, including physics, design, photo lithography, process technology and packaging, to name a few. Electronic systems of today often include a variety of analog, digital, wireless (RF), memory and interface functions, which are more and more supported by modern chip fabrication processes.

All integrated circuits, whether they are analog or digital, or a combination of these two, are built from transistors. Therefore, this chapter starts with a simple discussion on the operation of a transistor, because this forms the basis of understanding the properties of the various ICs built from them. The sizes of the transistors and interconnections on integrated circuits have been reduced continuously from several tens of microns (one micron = one millionth of a meter) in the 1960s to only ten nanometers (one nanometer = one billionth of a meter), today. In the same period, the supply voltage has reduced from 24 V to below 1 V. However, two equally sized nanometer transistors, running at only 1 V, do no longer show exactly equal behaviour. This difference in behaviour is called *mismatch*. *Matching* is particularly a requirement for analog circuits, whose performance is often based on the balance between the operations of two identical transistors. Digital circuits are much less affected by the increasing mismatch between transistors. The continuing scaling process following Moore's law is therefore much more beneficial to digital than to analog circuits. This is the reason why more and more analog functions are being transferred to the digital domain. As a result, most integrated circuits have much more transistors in the digital part than in the analog part and that ratio will certainly not reduce. This is also due to the fact that digital is getting closer and closer towards the system periphery, which was strictly analog domain before. However, still many analog circuits perform the operations that are needed in the growing number of wireless communications and data transfer systems. These systems use radio frequency (RF) communication.

© Springer International Publishing AG, part of Springer Nature 2019
H. Veendrick, *Bits on Chips*, https://doi.org/10.1007/978-3-319-76096-4_3

Although the title of this book suggests a digital focus of its contents, a discussion on analog and analog-to-digital functions will broaden the picture of the potential of integrated circuits. An additional summary of the basics of wireless communication should help the reader in understanding the complexity of advanced mixed-signal ICs that combine analog, digital, RF and memories on a single chip.

Many chip applications, today, include a variety of standard *intellectual property* (*IP*) *blocks* (*hardware*, e.g. processors, memories and interfaces). Although these blocks are completely identical on different chips, they may execute different programs (*software*) in order to execute the selected feature. Even such everyday applications as smartphones, tablets, cameras, radio, TV, GPS, etc. contain large portions of software. A small summary on hardware and software definitions and related system terminology will complete this chapter.

3.2   The MOS Transistor

Close to 90% of all integrated circuits are fabricated in a complementary metal oxide semiconductor (*CMOS*) technology, which features two types of transistors, an nMOS and a pMOS transistor. For the explanation of their operation, we will first focus on the nMOS transistor only, because the pMOS transistor operates fully complimentary.

Basically, a MOS transistor has three terminals: a *gate*, a *source* and a *drain* (Fig. 3.1). The gate is separated from the channel region by only an extremely thin insulating oxide. The thickness of this *gate oxide* is denoted by t_{ox}. The source and drain are made inside the wafer, while the thin white gate oxide, the gate itself and all transistor interconnections are made on top of the wafer.

The transistor operation is based on the amount of current that flows from drain to source. This current is controlled by the voltage on the gate. In fact, this current is defined by the amount of electrons that flow from source to drain. This flow of electrons can only be caused by connecting the transistor terminals to certain voltages. Every *MOS transistor* has a built-in threshold voltage (called V_t), below which it does not conduct (no flow of electrons). Only when the gate-to-source voltage

Fig. 3.1 Basic architecture of a MOS transistor

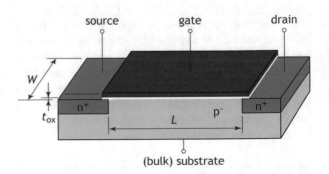

(V_{gs}) is larger than the threshold voltage that there will be a conducting channel, which is located below the gate right underneath the thin gate oxide and connects the source with the drain. In fact, the current flow through a transistor is similar to the water flow through a tap (Fig. 3.2). When the tap is closed, there will be no flow of water. A fully open tap will yield the maximum water flow. Let's assume the tap is closed. If we would need to turn the handle of the tap three times around before the water starts flowing, there will be an extremely small chance that the tap would leak when it is closed. However, if we need to give it only a quarter turn before the water starts flowing, there will be a big chance that the tap would leak when closed. The same holds for a transistor. The number of handle turns a tap needs before it opens is similar with the threshold voltage of a transistor. The lower the threshold voltage, the larger its leakage current when it's switched off. Since the supply voltage has reduced from 5 V to about 1 V over the last three decades, the threshold voltage was required to reduce almost proportionally, leading to relatively large leakage currents when the transistors are switched off. This causes leakage currents in smartphones, leading to power consumption even when they are in the standby mode. This is one of the reasons why their batteries need to be recharged every 1 or 2 days.

The width of the transistor channel in Fig. 3.2 is denoted by W, while its length is represented by L. It will be clear that a wider transistor channel allows more electrons to flow simultaneously from source to drain resulting in a larger current.

To integrate a maximum number of transistors onto a single chip, most transistors use the minimum possible channel length L defined by the technology.

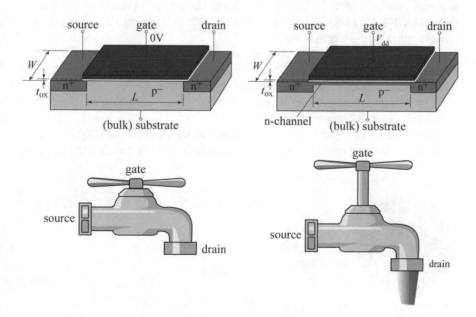

Fig. 3.2 Similarity between the operation of a MOS transistor and a tap

The process node is named after the channel length of the transistor: in a 60 nm CMOS technology, the target channel length is close to 60 nm, while in a 40 nm technology, it is close to 40 nm, etc. Technologies are developed for a fixed supply voltage V_{dd}, which is currently close to 1 V. This means that the transistor terminal voltages V_{gs} and V_{ds} can never be higher than V_{dd}. So, if we want to design a faster circuit in a given technology, we have to increase the transistor current by increasing the transistor width W. In analog circuits an increase in the transistor width may lead to a larger sensitivity of an amplifier, for example. So, optimising the widths of the various transistors that constitute a certain circuit 'gives' this circuit its required performance.

A lot more can be said about the basic transistor operation, but is not a requirement for further understanding of Part I. In Part II we will come back to the transistor operation, wherever necessary.

3.3   The World Is Analog

The word *analog* refers to a continuously changing process or perception. All signals that we feel, see or hear are analog by nature. You can feel pressure, you can hear sound, and you can see brightness. The intensity of this pressure, sound and brightness can vary over a wide range of continuous values. Other examples of physical parameters that are represented in an analog format are length, voltage, current, speed, temperature and time. Although we currently see a lot of *digital* applications like digital TV, DVD players, smartphones, tablet computers, e-readers, cameras, GPS, etc., they always need to interface with analog sensors, devices or circuits, because the world is analog.

Digital circuits operate with discrete (discontinuous) signal values, which make them perfectly suited for processing, storage and communication of huge amounts of binary data (digital representation of analog values).

A very simple example to explain the difference between analog and digital is in sound recording technology (Fig. 3.3).

The shape of the groove in a conventional analog vinyl record (Fig. 3.3 left) is directly proportional to the sound pattern. A turntable pick-up needle will physically

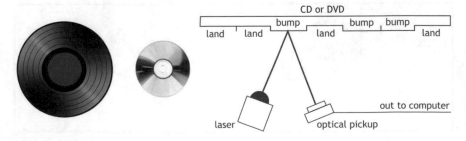

Fig. 3.3 Example of an analog (LP) and a digital (CD or DVD) music storage medium

follow the groove and moves the way a sound wave moves. Before storing music on a digital record, like a CD or DVD disc (Fig. 3.3 middle), it must first be converted to a digital format, which means that it is coded with 0's and 1's. A '0' is burned as a pit (bump) into the disc. A '1' is represented by the lack of a bump, which is called a land. The long series of pits and lands represent the digitally coded audio or video information stored on a disc as a continuous spiral of about 3–5 billion pits and lands. Playback can be done using a laser to scan the pits and lands (Fig. 3.3 right), which yield the original pattern of '0's and '1's again. Next, this pattern is converted back (decoded) to the analog format, which can then be amplified and listened to or viewed at.

To summarise, digital is discrete while analog is continuous.

In order to enable digital processing and storage of signals that are originally analog, these analog signals need to be converted to the digital domain. Analog values are usually represented in the decimal number system with base 10, meaning that the value of the digits is a power of ten and increasing with a factor of ten starting with the least significant digit (the most right digit). Digital values are represented in the binary number system with base two, meaning that the value of the digits is a power of two and increasing with a factor of two starting from the least significant digit. Therefore we will first explain how to convert from decimal numbers to binary numbers.

3.4 From Decimal to Binary Numbers

In the *decimal number system*, each digit represents a figure from 0 through 9. Figure 3.4 shows an example of a decimal and a binary number.

Originally the name *digit* comes from the Latin word *digital*, which means fingers. However, the value that a digit represents is also dependent on its position in the number. Each digit position has a different weight. The most right digit, which is also called the *least significant digit*, has a weight of 1 ($=10^0$), the second position a weight of 10 ($=10^1$), the third a weight of 100 ($=10^2$), etc. The decimal number 1367 is therefore the representation of $1\times1000 + 3\times100 + 6\times10 + 7\times1$.

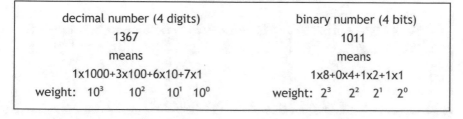

decimal number (4 digits)				binary number (4 bits)			
1367				1011			
means				means			
1x1000+3x100+6x10+7x1				1x8+0x4+1x2+1x1			
weight: 10^3	10^2	10^1	10^0	weight: 2^3	2^2	2^1	2^0

Fig. 3.4 Example of a decimal and a binary number

Digital circuits can only operate with binary numbers. The *binary number system* is less complex than the decimal number system. Binary means composed of two parts. So, a digit in this system can only represent two figures, 0 and 1; we also call it a *binary digit* or shortly: *bit*. Each bit position represents a value of a power of two. The most right bit, which is also called the *least significant bit* (*LSB*), has a weight of 1 (=2^0), the second position a weight of 2 (=2^1), the third a weight of 4 (=2^2), etc. The most left bit is called *most significant bit* (*MSB*). The binary number 1011 is therefore the representation of $1\times8 + 0\times4 + 1\times2 + 1\times1$, which is the equivalent of the value 11 in the decimal number system. So, we can translate every decimal number into a binary number. Figure 3.5 shows a table for the decimal to binary conversion for the decimal numbers 0 through 15.

A binary number is also called a *word*, and in this table a word contains four bits. A 4-bit word can represent 16 different decimal numbers, 0–15. A 5-bit binary word can represent 32 decimal numbers, 0–31, a 6-bit 64, etc. Suppose we want to translate the decimal number 1367 into a binary number. We then have to find the set of binary weights, whose sum equals the decimal number.

Figure 3.6 demonstrates the conversion process.

decimal	binary					decimal	binary			
	8	4	2	1	} weight		8	4	2	1
	2^3	2^2	2^1	2^0			2^3	2^2	2^1	2^0
0	0	0	0	0		8	1	0	0	0
1	0	0	0	1		9	1	0	0	1
2	0	0	1	0		10	1	0	1	0
3	0	0	1	1		11	1	0	1	1
4	0	1	0	0		12	1	1	0	0
5	0	1	0	1		13	1	1	0	1
6	0	1	1	0		14	1	1	1	0
7	0	1	1	1		15	1	1	1	1

most significant bit   least significant bit

Fig. 3.5 Decimal to binary conversion table

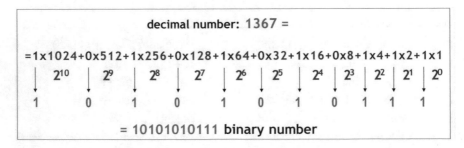

decimal number: 1367 =

= $1\times1024 + 0\times512 + 1\times256 + 0\times128 + 1\times64 + 0\times32 + 1\times16 + 0\times8 + 1\times4 + 1\times2 + 1\times1$

| 2^{10} | 2^9 | 2^8 | 2^7 | 2^6 | 2^5 | 2^4 | 2^3 | 2^2 | 2^1 | 2^0 |
| 1 | 0 | 1 | 0 | 1 | 0 | 1 | 0 | 1 | 1 | 1 |

= **10101010111 binary number**

Fig. 3.6 Conversion example from a decimal to a binary number

We first try to find the maximum power of 2 that is below the decimal value of 1367. This power of 2 is 1024 ($=2^{10}$). The next power of 2 is 512 ($=2^9$), but added to 1024, this would be more than 1367, and thus that power of two must not be added. The next power of 2 is 256 and added to 1024 results in 1280, which is still less than 1367, so this one must be added, etc. So far, we see that the decimal number 1367 = $1\times1024 + 0\times512 + 1\times256 + \ldots$, etc., as shown in the figure and thus equal to the binary number: 10101010111.

As explained before, digital circuits can only operate with digital signals, which are represented by binary numbers. Next, the analog signals must be converted to the digital domain before digital operations such as multiply, add, store and compare can be performed. The next subsection therefore discusses the conversion from analog to digital.

3.5 Analog to Digital Conversion

When we want to perform digital operations on an originally analog signal, we need to convert the analog signal to the digital domain. After the digital processing has been completed, we need to do the opposite. This will be explained on the basis of the digital audio system shown in Fig. 3.7.

Sound introduces local fluctuations in air pressure or vibrations of the air molecules which propagate through the air. Rapid vibrations create high tones (treble), while slow vibrations create low tones (bass). The diagram of Fig. 3.8 shows a sound waveform, which represents the fluctuations of the loudness of the sound over a certain period of time.

Larger fluctuations produce louder sound. The microphone is sensitive to the changes in air pressure and transfers these into electrical signals.

The output of the microphone (Fig. 3.7) is fed to an *analog-to-digital converter* (*ADC*), which quantises the analog signal at small but fixed instances in time. At each time instance, the ADC generates a digital value of the analog signal, which

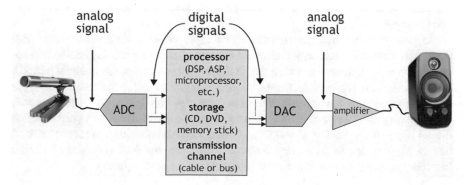

Fig. 3.7 Digital audio system

Fig. 3.8 Snapshot of a
sound waveform

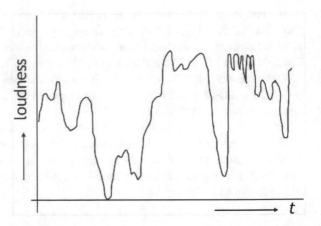

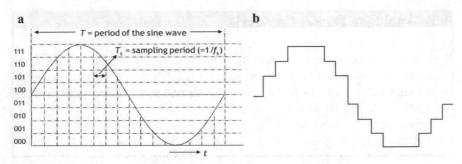

Fig. 3.9 (**a**) Analog sine wave signal and (**b**) its 3-bit digital representation

can then be processed or stored digitally. After that, the analog signal has to be
retrieved from the digital one, by a so-called digital-to-analog converter (DAC).
Next, the signal needs to be boosted by an amplifier. This amplified signal is then
converted back to sound by the loudspeaker.

We will now first focus on the function and operation of the ADC. Figure 3.9a
shows a sine wave representing an analog signal. The length of one full sine wave is
called the period T.

Suppose we want to *digitise* (create digital representation of) this waveform with
three bits. This allows us only eight discrete levels of the signal (Fig. 3.9b). T_s represents the repetition time at which the analog signal is sampled (the sample period).
It is clear that the higher the sample frequency (f_s), also called the *sample rate*, the
shorter the sample period and the more samples are taken per time interval.
Representing the sine wave form with only three bits does not create a very accurate
digital copy of the analog signal. For an accurate digital representation, we need
both a higher sample frequency and more bits. Figure 3.10 shows the relation
between the accuracy, the number of bits and the sample frequency.

The more bits we use in the digital word, the closer the digital copy matches with
the analog origin. During the 1970s, 4–8 bit words were used. The number grew in

Fig. 3.10 Relation between the accuracy of the digital representation, the number of bits and the sample frequency

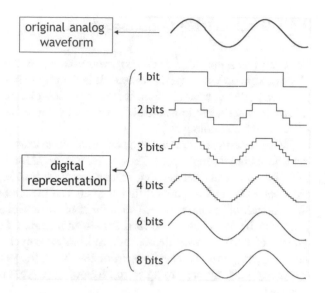

the 1980s and 1990s to 16 bits. Although the current resolution of ADC and DAC converters is limited to about 18 bits, addition and multiplication of such binary words may require up to 32–64-bit processors. Examples of this growing accuracy are the increased need for high-fidelity stereo audio, high-definition TV and more colours, etc. The combination of increasing number of bits and increasing sample rate (frequency) led to huge amounts of data (bits) that need to be timely processed by a (micro-) processor and stored in a memory.

More and more advanced electronic systems use wireless communication or data transmission, applying RF technology, which is the subject of the next subsection. The larger the number of bits per word, the more data must be processed, transferred and stored. This would require very complex processors, wide communication buses and extremely large memories. For this reason, very intelligent *data compression* techniques have been developed for a compact representation of audio and video signals. *JPEG* is a very popular technique used to compress the data (number of bits) that represent digital photographs and images. A commonly used technique for compressing digital audio is the well-known *MP3* format. Both techniques compress the amount of data close to a factor of ten. This is also very beneficial in emailing or up- and downloading music and pictures through the Internet. Playing 'MP3 music' requires that the CD/DVD player includes an MP3 decoder.

Much more could be said about the difference between analog and digital, but that is beyond the scope of this book. Analog and digital circuits and functions are discussed in Part II of this book.

3.6   Wireless (RF) Communication

Wireless communication (*wireless transmission*, over-the-air transmission) uses electromagnetic waves (*radio waves*). It is therefore also often referred to as *radio frequency* (*RF*) communication. It avoids the installation of billions of cables and allows easy access to inhospitable areas. Radio communication was the first application of RF technology.

When RF currents are sent to an antenna, electromagnetic waves are created that are transmitted through space. These waves propagate at the speed of light (*c*), about 300,000 kilometers per second, and can vary both in length and amplitude. The *wavelength* (λ) is the spatial difference between the start of one cycle of the wave and the start of the next cycle. The number of wave cycles per second is called the *frequency* (*f*) of the wave, which is expressed in hertz (Hz). So, a signal with a frequency of 1 MHz means that it has a million wave cycles per second. Wavelength and frequency are inversely proportional – when the wavelength goes down, the frequency goes up, according to the following expression: $c = f. \lambda$, because *c* is constant.

The total range of all frequencies, from 0 Hz to infinite, is called the *frequency spectrum*. This RF spectrum occupies several so-called frequency bands, which are assigned and coordinated by international standard bodies, according to Table 3.1.

Examples of wireless transmission include radio broadcasts (e.g. AM radio frequency, 535 kHz–1700 kHz, and FM radio frequency, 88 MHz–108 MHz), TV broadcasts (54 MHz–88 MHz, 174 MHz–216 MHz, 470–806 MHz), GPS (1200–1600 MHz) and cellular phones (800–1000 MHz, 1800–2000 MHz).

RF signals and systems have a certain *bandwidth*, which is expressed as the difference between the highest-frequency signal component and the lowest-frequency signal component. A typical analog voice transmission channel, as used by a phone, transfers human voice with a frequency range between 300 Hz and 3.3 kHz. It has a bandwidth of approximately 3 kHz. A TV channel has a bandwidth of approximately 6 MHz.

Table 3.1 RF frequency bands

Frequency band	Abbreviation	Frequency range	Wavelengths
Very low frequency	VLF	9–30 kHz	33–10 km
Low frequency	LF	30–300 kHz	10–1 km
Medium frequency	MF	0.3–3 MHz	1–0.1 km
High frequency	HF	3–30 MHz	100–10 m
Very high frequency	VHF	30–300 MHz	10–1 m
Ultra high frequency	UHF	0.3–3 GHz	1–0.1 m
Super high frequency	SHF	3–30 GHz	100–10 mm
Extremely high Freq	EHF	30–300 GHz	10–1 mm

RF Communication Systems

A basic communication system consists of three components:

- One or more mobile terminals, which act as interface between the user and the system. In case of a mobile phone, it can send and receive signals. In case of an FM radio, it can only receive.
- A *base station*, which acts as an interface between a wired network and the mobile (wireless) terminal(s).
- A network controller, which is connected to the host, from which it receives its commands and forwards these through the base station to the mobile terminal(s). It can operate as part of the base station or as a stand-alone device.

As shown above, there are many different wireless communication applications, which may not interfere with each other. Therefore, the low-frequency information signals are superimposed on a high-frequency *carrier signal*, also called carrier, to fit within the assigned frequency band of the application. This process is called *modulation*. Figure 3.11 shows two examples of such a modulation. In case of *amplitude modulation*, the original low-frequency signal varies the amplitude of the carrier signal, which results in the *AM* signal. With *frequency modulation*, the frequency of the carrier changes with the amplitude of the original signal, resulting in the *FM* signal. In fact, the information flies on the back of the carrier. This information signal can be an analog signal, but it can also contain digital data.

Figure 3.12 shows a wireless communication system. After modulation, the signal is amplified and sent to an antenna, which transmits radio waves (electromagnetic waves) into the surroundings. These waves are then received by the antenna at

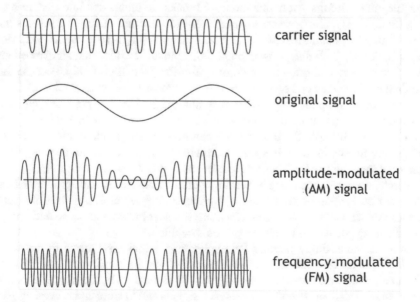

carrier signal

original signal

amplitude-modulated (AM) signal

frequency-modulated (FM) signal

Fig. 3.11 Examples of AM and FM modulation

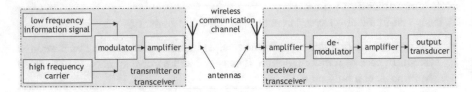

Fig. 3.12 Wireless communication system

the receiver, which is tuned to the carrier frequency to select the particular signal from the many simultaneously transmitted radio signals. In case of a radio, we tune the radio to the (carrier) frequency of the selected radio station. Then the small signal is amplified. After demodulation, in which the original information signal is retrieved again, another amplifier has to make the signal strong enough to drive a transducer, e.g. a loudspeaker or display.

We distinguish various communication systems. A *simplex communication* system is a radio system that allows one-way communication only: from transmitter to receiver, e.g. radio and analog TV. In a *half-duplex communication* system, both ends of the system can transmit and receive, but not at the same time, e.g. walkie-talkie. In a *full-duplex communication* system, both ends can simultaneously transmit and receive. Usually two different frequencies are then used: one for transmitting and one for receiving, e.g. satellite communication and cellular phones. A *transceiver* performs the two-way communication in these half- and full-duplex systems.

It is clear that RF communication requires huge amounts of electronic equipment and integrated circuits. More and more RF blocks are embedded into system-on-a-chip (SoC) designs for wireless applications, but these are mainly the small signal parts of an RF system. Most of the RF front-end parts (high-frequency circuits close to the antenna, including power amplifier, antenna switch, antenna impedance matching devices and often the low-noise amplifier) remain off chip, because their required performance can often not be met by standard CMOS technologies.

A few wireless applications will now be summarised. A very popular application is *wireless Internet*. It uses the communication standard *WiFi* in a *wireless local area network* (*WLAN*). Such a network can be used in private homes but also in offices and hotels. Next to PCs and smartphones, this network can also be used to connect consumer products such as televisions and DVD players.

Wireless communication is increasingly used by devices in and around the body. They can be located under the skin (*implanted devices*) or along with objects or clothes (smart watches, *wearable electronics*, *smart textiles*) to sense and analyse biological signals and to monitor a person's health.

Another increasingly popular RF application domain is formed by the *RFID* (*radio frequency identification*) cards, tags and chips. All RFID cards and tags contain an antenna to transmit and receive the RF signals and at least one integrated circuit to modulate or demodulate these RF signals and to store and process the data and/or perform other dedicated functions. There are two different categories of tags:

Fig. 3.13 A few RFID applications

passive and active. *Passive RFID* tags are relatively cheap and do not have a transmitter. They extract their energy from the RF signals sent by the card reader and can only respond to these signals. They can usually only be read over short distances: from a few centimeters to several meters. *Active RFID tags* use a battery (or alternative power sources, such as solar power) for their power supply and are commonly developed for long-range communication with reading distances up to 100 m.

RFID tags operate at various frequencies, which depend on the application, on the environment (e.g. within the air, within a pallet, in cargo containers or on a metal) and on the reading distance (e.g. centimetres or metres). RFID chips are widely used in identification applications (personal identification, electronic passports and medical, product and animal identification), in security, in product tracking, in tagging and in sensing applications (temperature, tyre pressure). Figure 3.13 shows a few examples of RFID applications: a car key, an electronic passport and a contactless credit card.

Future market potentials of RFID are huge. In 2016, 50% of the world's top 100 retailers are using RFID in their stores to track stock levels of items and their locations. It is expected that in the near future, the barcodes on many other products will also be replaced by cheap RFID chips. It greatly improves the logistics and inventory control and also supports automatic product tracking, electronic price reading and automated cash registers. In clothing sales, for instance, RFID tags can also communicate information about these clothes, such as size and colour. RFID is also increasingly used for electronic toll collection to eliminate the delay at toll bridges and toll roads.

While it is already normal to use RFID tags for animal identification, many people are still sceptic about their use for human identification and medical purposes, because of privacy rights and the possibility of identity theft. These arguments prevent many RFID application initiatives from moving forward at the same pace that technology would enable them.

Near-field communication (*NFC*) is another smart form of RFID. It uses 13.56 MHz contactless bidirectional communication and can transfer a maximum of 424 Kb/s between electronic devices over distances less than 10 cm. Information can be simply exchanged by holding the electronic device close to a variety of smart tags. An NFC-enabled smartphone will allow you to do payments, buy train tickets, grab information from public billboards, make connection to a friend's phone, etc. Some smartphones use NFC technology to enable direct data transfer (pictures, music, contacts, etc.) by holding two phones together.

Another example of the use of wireless transmission is *Bluetooth* communication. It is especially developed for creating personal communication networks with a maximum range of about 10 m. Its function is to replace cables between portable and/or fixed electronic devices, such as mobile phones, handhelds, walkie-talkies, headsets and laptops and between a PC and its peripherals, such as keyboard, mouse, monitor and printer. Bluetooth devices operate with frequencies between 2.402 GHz and 2.480 GHz and can automatically detect and communicate with other Bluetooth devices without any user input. This frequency range is specifically reserved by international agreement for industrial, medical and scientific (ISM) devices. To avoid interference with other networks, Bluetooth devices only consume about 1 mW of power during transmitting its signals. Because of this and because a Bluetooth chip is cheap (only a few US$), it has a huge potential for many application areas, such as office, home, medical, e-commerce, traffic, etc.

Finally, as science progresses, alternative wireless technologies may soon replace existing ones. An example is the so-called *WiFi Direct* technology, which also allows electronic devices to directly communicate with each other without the use of a base station. It can operate faster and across greater distances than Bluetooth. Although consuming much more power, it is its nearest competitor and is expected to replace Bluetooth in certain application areas soon.

The above examples showed that wireless communication already exists for more than a century and that it is being used in an increasing variety of applications. It enables the *Internet of things* (*IoT*), which usually refers to smart cities, smart homes and smart buildings, for example. IoT is a system that enables real-time communication between connected computing, mechanical, analog, digital and memory devices, objects, animals or people, each provided with a unique identifier. This enables autonomous data transfer over the Internet, without requiring human-controlled interaction. It has a tremendous impact on our society and makes the world smaller. Moreover, it still has a huge potential for many new, yet unknown, future applications.

3.7 Mixed-Signal Circuits

As their name implies, mixed-signal circuits combine a mix of analog, digital and RF circuits onto the same piece of silicon. Digital circuits operate on signals with only two discrete voltage levels: ground level, corresponding to the logic '0' state, and supply level, corresponding to the logic '1' state. Before a digital circuit can switch from one logic state to the other, its input signal(s) must have passed a certain threshold voltage, which lies in between these two discrete levels. Digital circuits are therefore less prone to noise and distortion. A disadvantage of digital circuits is that most of them operate synchronously, which means that the data processing is controlled by one or more clock signals. As a result, many signals within the digital parts of a chip switch simultaneously, which is why they are a major contributor to the noise in mixed-signal ICs.

Particularly for the analog and RF circuits, the scaling of the physical sizes and of the supply voltage has a negative influence on their performance and robustness of operation. Analog signals usually have low amplitudes and carry the information in their form and shape. In electronic circuits, these signals are subjected to noise, either introduced by neighbouring signals (*crosstalk*) or by the uncontrolled thermal movements of electrons (*thermal noise*) that temporarily change the current. They therefore require a large *signal-to-noise ratio* (*SNR*). Scaling of the physical sizes to create denser circuits causes the digital and analog circuits to coexist in close proximity. This will increase the interference of the digital signals into the analog circuits, which may change the shape of their signals. Reducing the supply voltage would reduce their amplitude as well. Both effects impact the design complexity of mixed-signal circuits, which will become an increasingly time-consuming task.

RF signals usually operate at specific frequency ranges and often require circuits with odd-shaped and hybrid devices on the printed circuit board (PCB). Antenna products, such as mobile devices, can both send and receive noise signals and require additional RF shielding. This may be included in their package or mounted onto the printed PCB.

Today's advanced CMOS technologies enable the integration of a complete *system on a chip* (*SoC*). It is therefore increasingly likely that these chips will contain analog, RF, digital, memory and interface circuits. This poses huge challenges to both the IC designers and the computer-aided design (CAD) tool developers. The nature and operation of analog and RF circuits do not allow design automation to the level that digital and memory circuit designs do. The combination of all four of them on one chip increases this challenge. Next to the design complexity, also the testing of a mixed-signal IC is a complex, time-consuming and expensive task.

To summarise, mixed-signal ICs are far more complex to design and test than purely digital ICs. In many cases, however, these additional IC development costs are offset by a better average price setting, because *analog intellectual property* (*analog IP*) usually has a higher commercial value than *digital IP*.

3.8 Memories

As was already discussed in Chap. 1, close to 30% of the total chip market is in stand-alone memory chips. These chips almost contain 100% of their transistors in the memory cell. On almost all other chips, often more than 80% of their transistors are also in an (embedded) memory. So, of all transistors produced in the world, today, more than 99.5% end up in a memory. Therefore, memories are extremely important for electronic systems and require the utmost of the lithographic and fabrication tools in a semiconductor fab.

Fig. 3.14 DRAM memory module

Memories are used to store the results from operations performed by digital processors. Applications, like a PC, require extremely fast storage (writing) and reading operations of their memory. Usually the information stored in such a *random-access memory (RAM)* is rapidly changing over time. This type of memory is also called a *volatile memory* because it will lose its data when the supply voltage is switched off. To allow the processor to have a fast read and write access to the memory, it is often embedded with the processor on the same chip. Next to such an *embedded memory*, a RAM is also used to be plugged into a memory bank on the computer's mother board. These *stand-alone memories* usually have a much larger memory capacity (more gigabits per chip) than an embedded memory. They are most commonly implemented as *dynamic random-access memory (DRAM)*. Such a memory consists of very small memory cells, which each uses only one transistor and capacitor. Most PCs and laptops have their memory chips organised in memory banks. A single memory bank may contain any number of memory chips. If we assume that each of the memory chips on the *memory bank*, or memory module, in Fig. 3.14 contains 4 GB of memory, then the memory bank contains 8 GB. This memory bank is plugged into an array of sockets (*memory slots*) on a printed circuit or motherboard (at the bottom-right side in Fig. 3.15). PC motherboards can have two or more memory banks and often have a few empty extra sockets (slots) available to enable the plug-in of additional memory to upgrade the computer after a couple of years.

Figure 3.15 shows an example of a bare naked motherboard. The processor, interface and memory chips are not yet plugged into this board. The four memory slots are clearly visible in the right-lower corner.

While RAMs are used to store intermediate data and results, ROMs are used to store permanent data or fixed programs. A *read-only memory (ROM)* is a *non-volatile memory*, which does not lose its data when the supply voltage is switched off. It can only be read, because the program is stored during the fabrication of the memory.

Many applications, however, require a non-volatile memory in which the data needs to be changed once and a while. Applications of such a non-volatile memory include digital cameras and e-readers (SD card, compact flash card), MP3 players, smartphones (SIM card), tablets, laptops and memory sticks. Figure 3.16 shows a couple of these non-volatile memory applications.

When taking a picture, the picture should remain on the memory card in case the camera is switched off. However, it is possible to erase the picture and capture and

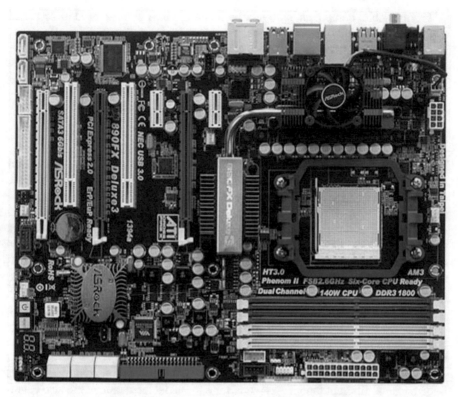

Fig. 3.15 Empty motherboard showing various slots for processor, interface and memory chips. (Photo: AMD)

Fig. 3.16 Various examples of non-volatile memory applications; from left to right: compact flash card, SD card, SIM card and four USB memory sticks

store a new one. In an MP3 player, songs must remain in the memory until they are replaced by other songs. The subscriber identity module card (*SIM card*) used in some mobile phones must maintain such information as personal identity information, phone book and text messages and other data. The SIM card can therefore be easily plugged into a new phone while taking all the stored information with it.

Advanced memory cards and memory sticks can store many gigabytes of information and allow the storage of thousands of pictures and/or MP3 songs.

In most tablet computers, non-volatile memory chips are also used as so-called solid-state drive (SSD), which replaces the magnetic *hard disc drive* (*HDD*).

Much more information on memories is available from Chap. 6 in Part II.

3.9   Hardware and Software, Architecture, Instructions and Algorithms

Most complex electronic devices and systems, today, contain both hardware and software. The term *hardware* refers to all physical components from which these devices and systems are built. Examples of hardware in a PC are motherboard, processor, hard disc and DVD player/burner but also display, mouse and keyboard and their interconnections. Examples of hardware on a chip are individual transistors and complete functions such as multiplier, adder, filter, processor, memory, interconnections and interfaces. The *architecture* of the chip is the hardware concept, which is designed such that it can execute the required functions while meeting the specifications in terms of performance and cost.

When an electronic device can only execute one single function, it is called a *dedicated device*. This means that when it is switched on, it always knows exactly what it should do and it would continuously repeat it. Examples are MP3 player, video processor, digital camera, GPS, etc.

A *general-purpose device*, like a programmable processor, is built to execute a wide range of different functions. A smartphone and tablet computer are also examples of a general-purpose device. A smartphone includes hardware to function as a mobile phone, to play MP3 music or to navigate using the built-in GPS receiver. Next to that is it has a camera, radio and several other hardware functions. However, since the introduction of the first smartphone in 2007, millions of applications (*apps*) have been developed to build features around its integrated hardware functions. In fact, an application is nothing more than a program that runs on the device, which can then perform a dedicated feature. An *app* may turn a smartphone into an organiser, into a scientific calculator, into a TV screen, into a chessboard, into a city map, into a spread sheet, into a compass, into a music library, into a music recogniser, etc. Many of these apps, which can be downloaded from Internet stores, are free; others cost a few dollars each. These *programs* are also referred to as *application software*, which includes everything that determines what the hardware should do. This is quite different from *system software*, which controls the operating system and determines how the hardware does it.

The *operating system* performs such basic tasks as reading input from the mouse and keyboard, controlling the disc storage (managing files and directories) and controlling the peripheral devices like display, printers, external hard discs, etc. It also has to arrange the various windows on the display and must prevent interference between two simultaneously executed programs. Finally it should check access authorization by verifying usernames and passwords and keys.

The *program* contains a sequence of operating steps (*instructions*), which exactly tell the system what to do. Most programs use algorithms. An *algorithm* is a kind of recipe for an effective execution of the program. It usually consists of a sequence of well-defined basic operations, such as multiply two numbers, add this product to the previous result and store the final result into a memory. An algorithm must take care of an efficient execution of the instructions by the hardware.

Because video processing is quite different from audio processing, which on its turn is quite different from telecommunication, a general-purpose processor cannot execute all applications evenly efficient, although its program may use algorithms that are optimised for each of the application domains. To achieve the best performance, each application domain requires its own optimised dedicated processor architectures, algorithms and programs. Most PCs therefore include, next to the general-purpose (micro-)processor, dedicated graphics and/or video processors to increase the performance when running graphics or video applications, respectively.

3.10   Conclusion

Digital circuits only operate on binary numbers. Therefore all decimal values need to be converted to binary numbers. Because digital circuits are very robust and can be realised with continuously scaling feature sizes, more and more signal processing that was originally done in the analog domain has been and is still being moved towards the digital domain. Analog-to-digital and digital-to-analog converters therefore play an important role in the overall system performance.

An IC is usually part of the system hardware and may contain analog, digital, RF, mixed-signal and memory circuits. An overview of the basic concepts behind these circuits can only present a flavour of the potentials of these circuits.

Hardware usually refers to the physical components of the system, while software most commonly refers to the program that must be executed by the system. This chapter included some related definitions and terminology, to show that in most electronic systems, hardware and software are jointly executing the required tasks to perform the specific electronic function.

Much of the discussions in this chapter will also enhance the readability of many of the topics discussed in Part II.

Other Interesting Stuff
There is a lot of interesting material on the basics of analog, RF, digital, memories, hardware and software available on the Internet. In many cases a Google search 'introduction to...' or 'white paper...' will yield many hits covering the topics at various levels of detail. A selection is not given here, because it depends very much on the background and interest of the reader. The reader is advised to scan the chapters in Part II as well, since they also contain more details on these subjects.

Chapter 4
The Chip Development Cycle

4.1 Introduction

In this section we will present an overview of the basic steps that are required to create a state-of-the-art chip today. Figure 4.1 illustrates these steps. Because about 90% of all ICs are made in a complementary metal oxide semiconductor (*CMOS*) process, this section, but also the remaining of the book, will be focussed on this technology only. For the following discussions, the reader is advised to regularly refer to the figure below.

4.2 Design

The continuing development of IC technology during the last couple of decades has led to a considerable increase in the number of devices per unit chip area. The resulting feasible IC complexity currently allows the integration of a complete *system on a chip* (*SoC*), which may comprise hundreds of millions to several billion transistors. Consequently, the design of such chips no longer simply consists of the assembly of a large number of logic gates. This poses a problem at a high level of design: how to manage the design complexity. Besides this, the growing influence of parasitic and scaling effects, such as noise and interference, may reduce chip performance dramatically and requires a lot of additional design resources to take and implement adequate measures. Such ICs combine signal processing capacity with microprocessor or microcontroller cores and memories. The dedicated signal processing parts take care of the computing power (workhorse), while the microprocessor or controller serves to control the process and possibly performs some low performance computation as well. The memories may store program code and data samples. Finally, since the world is analog, most ICs also contain on-chip analog interface and pre- and post-processing circuits as well as an increasing number of wireless interfaces. The development of such heterogeneous systems on one or

© Springer International Publishing AG, part of Springer Nature 2019
H. Veendrick, *Bits on Chips*, https://doi.org/10.1007/978-3-319-76096-4_4

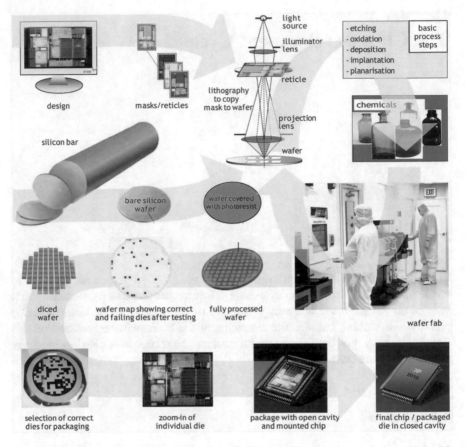

Fig. 4.1 Overview of the creation cycle of a chip: from design to packaged device. (Wafer fab photo: Core Systems, Inc)

more ICs, for instance, may require tens to even hundreds of man-years, depending on their complexity. Microprocessors for standard PCs, laptops and servers, e.g. the Intel and AMD processors, may even require several thousand man-years of development time.

Before starting a chip design, the complete functionality of the chip must have been specified according to the requirements of the intended application. This is done by system architects, who are not only familiar with the application but who are also capable of creating the best architecture for the chip. This means that they must specify the blocks (*cores*) (e.g. microprocessors, multipliers, adders, digital signal processors (DSPs), memories, analog and RF functions, interfaces, etc.) that are needed on the chip to guarantee an optimal execution of the required functionality. Next to this, the chip must perform in the most efficient way, in terms of speed, power consumption and area usage. Some of these cores may be available from existing libraries, because they have already been used before on other chips. The use of in-house or externally developed IP cores is generally referred to as *reuse*. The cores that are not yet available must then be designed by IC designers, who are also responsible for the final floor planning and full integration of all the blocks (cores) on the chip. So, a complete chip design consists of a combination of reused existing cores and newly designed cores. These latter cores, which need to be designed from scratch, are developed by using complex tools, which can map certain core functionality onto a library of standard cells in a given CMOS process. The reader is advised to read Sect. 5.5 in PART 2 on digital core and chip design, to get an impression of design methods used to create individual cores (e.g. multipliers, microprocessors, etc.) and Sect. 7.5 in Part II on how to integrate these cores via floor planning into a complete chip design.

More details on the design of memories, analog, interface and mixed-signal designs can be found in Part II, Chaps. 6 and 7.

Once the design is verified by simulations and is expected to function correctly, the final *design database* is created.

A chip is fabricated by processing various layers inside and on top of the wafer. The transistor sources, drains and channels are fabricated inside the wafer, while its gates (see Figs. 3.1 and 3.2) are created on top of the wafer surface. A chip is built from hundreds of millions to several hundred billion transistors, connected through wires in many different metal layers, which are stacked on top of each other and above the transistors. The design database must therefore contain the full description of the layout of the chip, which includes all individual geometric patterns in each of the layers. During chip fabrication, particularly during the lithographic steps, the patterns in these layers are then copied onto the wafer to create the transistors and metal interconnections, layer by layer.

4.3   Substrates (Wafers)

Almost all integrated circuits are fabricated on *silicon* substrates. Silicon is extracted from the mined silicon dioxide or shortly silica, which is largely available as beach sand, sand stone and quartz and thus relatively cheap. It offers the possibility to change the mobility of the charge carriers, which are responsible for the transistor currents and thus for the circuit speed. It easily oxidises to *silicon dioxide*, which is a good insulator that is used to isolate transistors from each other as well as to isolate the metal interconnect layers from each other and from the transistors. Moreover, after all layers have been processed, a silicon-dioxide (or nitride) *passivation layer* on top of the fabricated chip protects it against humidity, corrosion and contamination. These are some of the reasons why silicon has surpassed germanium already five decades ago, as the material of choice, and why most integrated circuits are fabricated on silicon substrates. Because these substrates are usually thin slices of silicon material, they are called *wafers*.

To create silicon wafers, first pure silicon is heated at temperatures up to 1500 °C in a huge furnace. Then a seed of single silicon crystal is mounted on a shaft and is dipped into the molten silicon. This seed is then slowly rotated and raised upwards out of the melt just fast enough to pull the molten silicon with it by cohesion, thereby physically growing the silicon crystal. In this so-called Czochralski process, the *crystal growth* is a continuous process of forming new thin films of the silicon melt on the bottom of the cooled previous films, roughly at about 20 mm an hour. The diameter of the grown *monocrystalline silicon bar*, also called an *ingot*, varies over the length, and a grinder is used to create a bar (Fig. 4.2) with a homogeneous diameter, which can vary from several inches up to 12 inch ($\approx$300 mm). Next, wafers are sawn by a diamond-coated saw. Because the transistors are fabricated close to the silicon surface, their performance and reliability are very much dependent on the flatness and crystal integrity of the silicon wafer surface. Theoretically, for good MOS transistor operation, the wafers could be as thin as a micron, but with this thickness, a wafer would easily break during handling. Therefore most wafers have a thickness between 400 µm and 1 mm.

Four decades ago, the average die size was close to 10 mm^2. A 1-inch wafer, at that time, could still contain more than 100 dies. As the complexity and sizes of the chip grew, so did the wafer size in order to still carry a sufficiently large number of dies for a commercially attractive chip fabrication. Figure 4.3 shows how the wafer size has evolved from 1 inch in the 1970s to the 12 inch today.

To complete the wafer manufacturing process, the wafer is flattened and planarised through a couple of etching and polishing steps. Now the wafer is ready to be used in the chip fabrication process. (More details on substrates and wafers can be found in Part II, Chap. 8.)

Fig. 4.2 Silicon bar with a few sawn wafers

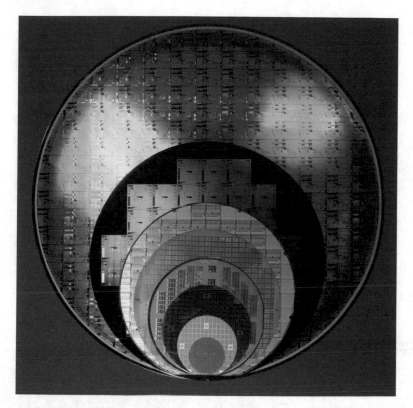

Fig. 4.3 Increase of the wafer size from 1 inch in the 1970s through 2, 3, 4, 5, 6

4.4   Fabrication

As discussed before, an integrated circuit (chip) is built from different layers. The pattern in each layer is represented by a pattern in the corresponding *mask* (*reticle*) as shown in Fig. 4.4. Complex chips are built from many layers and may require more than 60 masks. Each layer requires several production steps, which may total up to more than 1000 for an advanced chip. However, before executing a production step, the corresponding mask pattern must be copied onto the wafer to define the areas that must be affected by this step and the areas that must not.

So, a complete chip design, which is called a *database*, contains the complete information of all individual geographical shapes in every mask. Eventually, the patterns in each mask have to be copied onto the wafer during the *photo lithography* process, which precedes the individual manufacturing process steps.

The sequence of lithographic steps is explained with the aid of Fig. 4.5. We start our discussions somewhere during the fabrication process. We assume that several layers have already been processed and patterned and that we are now at a stage that we need to copy the next mask image onto the wafer to form the pattern in the next to-be-processed layer (Fig. 4.5a).

Fig. 4.4 A 4× reduction reticle for step and scan systems. (Photo: ST Microelectronics)

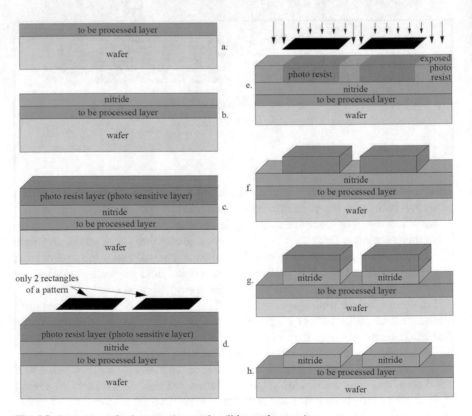

Fig. 4.5 Pattern transfer from mask to wafer. (lithography steps)

Usually, the first step is the deposition of a protective nitride layer (Fig. 4.5b). Subsequently, this nitride layer is covered with a photoresist layer (Fig. 4.5c). A mask is used to selectively expose the photoresist layer to light (Fig. 4.5d and e). Openings in the mask correspond with the exposed areas on the wafer. Exposure to light changes the chemical properties of the photoresist in the exposed areas. That photoresist is then developed and removed in the exposed areas. The resulting pattern in the resist after development (Fig. 4.5f) is a copy of the mask pattern and acts as an etch barrier in the subsequent nitride etching step (Fig. 4.5g), in which the unprotected nitride regions are removed (stripped). Finally, the remaining resist is removed, and an image of the mask pattern remains in the nitride layer (Fig. 4.5h). This nitride pattern acts as a barrier for a subsequent processing step.

Now the real fabrication process step (etching, oxidation or implantation) is performed through the open patterns in the nitride layer. We will describe a few basic *CMOS* process steps in connection with Fig. 4.6. CMOS stands for complementary metal oxide semiconductors. Complementary means that the circuits are built from two transistors, which are each other's complement, both in their fabrication and operation, as explained in Part II, Sect. 8.3. The abbreviation *MOS* refers to the cross section of the transistor as shown at the right side in Fig. 4.6.

We start with the situation of a bare silicon wafer on which the first nitride and photoresist layer are deposited (Fig. 4.5c). So, the to-be-processed layer in this case is the original silicon wafer. Then, the pattern in the first mask is used to define the open areas in the nitride layer through which silicon must be etched away. This creates trenches in the silicon wafer. After this etching step, the nitride layer is removed. These trenches are then filled with oxide and form the so-called STI isolation areas between the transistors (Fig. 4.6). Next a very thin gate oxide isolation layer, which is between 1 and 3 nm thick, is deposited on the wafer. This can currently be done with atomic accuracy. This step is followed by the deposition of a polysilicon layer, which will act as the gate material. Then a new nitride and photoresist layer are deposited on top of this (red) polysilicon layer.

The second mask is used for patterning this new nitride layer by the second photolithographic step. The next process step, e.g. etching of the deposited polysilicon layer, is then performed through the openings in this nitride layer using etching

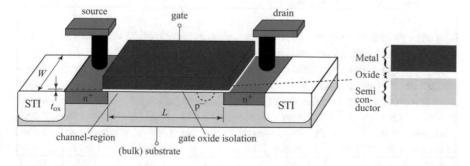

Fig. 4.6 Architecture of a MOS transistor

chemicals that only etch the polysilicon away. On the locations where the polysilicon remains, it acts as the gate of the transistor (Fig. 4.6). Then these nitride and photoresist layers are removed again, and a new nitride and photoresist layer are deposited, preparing for patterning the wafer for the next process step, etc. The lithography steps, as shown in Fig. 4.5, are thus repeated for every new mask, followed by the subsequent fabrication steps that are performed through the openings in the corresponding nitride pattern (copy of the respective mask pattern). The yellow circular arrow in the right part of Fig. 4.1 reflects this sequence of lithography and fabrication steps, which are repeated until all masks are used and all required process steps have been performed. The total number of process steps for state-of-the-art processes today is between 1000 and 1500.

Figure 4.7 shows an on-scale cross section of two transistors and various metal interconnection layers. The transistors are made inside the top of the wafer, which is the lowest layer. The contact layer contains contacts at those locations where a transistor source, drain or gate must be connected to the first metal layer. The connections between two metal layers are implemented by contacts in the respective via layers. For instance, a contact between the first and second metal layer is created in the first via layer. The second via layer contains all contacts between the second and third metal layer, etc. Some transistors are connected to nearby transistors. Then these signals only flow through the lowest two or three metal and via layers. Connections between transistors that are far away from each other may even use metal tracks in four or five different metal and via layers. The lower metal and via layers are relatively thin, because they are mostly used to transfer the signals through the chip over relatively short distances. The intermediate metal layers, today, are a little thicker and used for signals that need to cover longer distances. The top metal and via layers are usually much thicker, because they need to distribute the large supply currents through the chip.

The fully processed wafer now contains an array of chips (see Fig. 4.1). As long as these chips are on the wafer, these are also called *dies*. (More on lithography and fabrication processes can be found in Part II, Chaps. 9 and 10, respectively.)

4.5   Testing and Packaging

The extremely small physical details on a chip in combination with very tight margins in electrical operation require extremely high standards of reproducibility of the lithographic and fabrication tools and processes. This does not prevent that a certain amount of dies on the wafer does not function correctly. Therefore all dies are individually tested on the wafer before they are packaged. To enable automatic testing of the dies and allow wire bonding, all input, output and power supply connections are usually located at the periphery of the die. These connections, called bond pads, are relatively large, e.g. 30 μm by 30 μm. Figure 7.6 in Part II shows an example of a chip with the bond pads in its periphery. This electrical testing, which is often referred to as probing, is done by means of metal needles (probes) on a

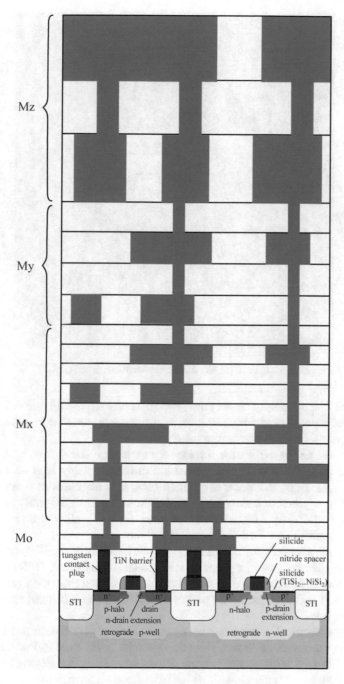

Fig. 4.7 Cross section of a fully processed chip, showing the various layers from which it is built

Fig. 4.8 Example of a probe card. (Photo: MICRAM Microelectronic GmbH)

probe card that physically contact the bond pads on each die. Figure 4.8 shows an example of a probe card with 120 pins. Many different tests are required to completely verify a correct operation of the chip.

This *e-sort test* or *wafer-sort test* identifies which of the dies operate correctly and which do not. The results of the test are stored in a *wafer map* on a computer (see Fig. 4.1). Next, the wafer is grinded back to a thickness of a few hundred microns and attached to a soft elastic carrier substrate. Then the wafer is separated into individual dies by a diamond saw or *laser dicing*. To allow this dicing, *scribe lanes* of 50–200 µm are designed around each die.

Finally, the die needs to be packaged to allow handling of the IC for mounting it onto the application board, e.g. inside a PC, smartphone, tablet, GPS navigator or TV. The correct functioning dies are automatically picked from the flexible carrier (see Fig. 4.1) and attached to the package substrate using a thermal and electrical conducting adhesive.

Next, the electrical connections between the bond pads of the die and the pins of the package are realised through soldering a wire (*wire bonding*) or by *flip-chip bonding*. With the latter technique, the chip is assembled face down directly onto solder bumps on the circuit board, which have the same formation as the bond pads on the die.

A package also serves as a mechanical and electrical interface between the die and the application board, as well as protection against environmental mechanical

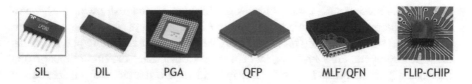

Fig. 4.9 Various package images

and chemical attacks. Finally a package improves the heat transport between the IC and the environment.

Figure 4.9 shows a variety of different packages.

The single in-line (SIL), dual in-line (DIL) and pin grid array (PGA) packages are so-called through-hole packages. When they are placed on a board, their pins are physically put through holes in the board. The quad flat pack (QFP) and micro lead frame (MLF) packages are *surface-mount packages* whose terminals are mounted directly on the board terminals. These two package categories, as well as the flip-chip bonding technology, allow smaller packages, shorter wires and a better electrical and thermal connection to the board.

Figure 2.1 (Part I) shows an example in which surface-mounted chips are attached to a printed circuit board (PCB). In Fig. 2.4 several packaged chips are soldered to a small substrate as used in a smartphone. Because many applications require an extreme density of the electronic components, more and more so-called *3D stacking* techniques are being used, in which various dies or packaged chips are stacked on top of each other. Part II, Figs. 14.10, 14.11 and 14.12 show a couple of examples of such stacked dies and stacked packages. Finally, Part II, Fig. 14.8 shows the package trends over more than five decades.

More test and package details are discussed in Part II, Chaps. 12 and 14, respectively.

4.6 Conclusion

This chapter presented only a flavour of the difficult tasks and processes involved in the complete development cycle of a chip. It concludes the introductory part of this book: Part I. Hopefully this part has brought the reader a clear picture about the large variety of existing electronic devices and systems that use integrated circuits. And, what's more, it should have raised awareness about the complexity of all involved semiconductor disciplines: design, substrates, fabrication, packaging and test.

In Part II these topics are explained in a more detailed way. A lot of effort has been devoted to presenting the various topics in an explanatory way, by including lots of graphical material and photographs so that the readers can digest the information more easily. However, it must be stated that the subject of *microchips and bits* remains a technical subject and that some affiliation with technology in general would help the reader to understand things faster. For the readers that are less interested in more details about chip design, fabrication and test, I would advise to still read Chaps. 7 and 15 in Part II. Chapter 7 is important because memories form a major part of the total chip market and are essential parts in mobile phones, cameras, MP3 players, GPS systems, etc. Chapter 15 discusses the limitations of further scaling of integrated circuits and introduces the reader into the incredible potentials of combining microelectronics (nanoelectronics) with nanotechnology. This will result in the creation of a large range of new devices, sensors and actuators. Particularly the introduction of these devices in biotechnological and biomedical applications will create an enormous potential of compact electronic devices and systems that will impact human life, with respect to comfort and health beyond today's imagination.

Part II
A Deeper Dive

Introduction

Integrated circuits (chips) are increasingly used to collect, process, store and transfer all types of data, make decisions and provide automation and control functions to deliver a significant contribution to the safety, health and wellness of mankind.

Part I is specifically developed for people with almost no or little technical knowledge. It presents an overview of the electronic evolution and discusses the similarity between a chip floor plan and a city plan, using metaphors to help explain concepts. It includes a summary of the chip development cycle, some basic definitions and a variety of applications that use integrated circuits. This part can also be used as a step-up to Part II.

This part (Part II) digs deeper into the details. Although this part is perfectly suited for professionals working in one of the semiconductor disciplines who want to broaden their semiconductor horizon, much of it is also digestible by people with only little technical background. For all categories of readers interested in *Bits on Chips*, I can only advise: scan through the subjects and find out what's of interest to you.

Chapter 5 focuses on the so-called digital IP, which includes all types of digital circuits. It starts with a discussion on the basic logic gates and where they are used for. After a straightforward implementation of adders and multipliers, the chapter continues with a description of the design, implementation and verification of more complex digital blocks on a chip.

Stand-alone memories (mostly DRAMs and flash memories) currently represent 25% to 30% of the total semiconductor market revenue. Moreover, also in logic and microprocessor ICs, embedded memories represent close to 80% of the total transistor count. So, of all transistors produced in the world today, more than 99.5% (!!) end up in either a stand-alone memory or in an embedded memory. This share is expected to stay at this level or may even increase, since the need for data storage is insatiable. The majority of available memory types are therefore examined in Chap. 6.

Their basic structures, operating principles and characteristics are extensively explained. In addition, the relationship between their respective properties and application areas is made clear.

Because the world is analog, many of the IC's input and output signals are analog, which require analog processing and analog-to-digital conversion before the signals can be digitally processed and/or stored. This means that on-chip analog circuits can often be found close to the interfaces and input/output (I/O) terminals of the chip. For this reason, Chap. 7 combines discussions on analog circuits and IP with the chip interfaces and I/Os.

The behaviour, performance and reliability of integrated circuits are very much dependent on the substrate on which they are built. Most integrated circuits are fabricated in CMOS technology on silicon wafers. The characteristics and properties of various wafer substrates are discussed in Chap. 8, based on their atomic structures. It also describes how the properties of these materials (resistivity, conductivity, leakage currents) can be manipulated by locally implanting atoms of different materials into the silicon wafer to create nMOS and pMOS transistors or to improve their performance.

When the design (layout) of a chip is completed and verified, it contains a geometric description of all the layers from which the chip will be built in the wafer fab. The layout patterns in each of these layers are copied from the geometric description into one or more masks. Advanced CMOS chips, today, may require more than 60 masks to define all necessary layers and processing steps. The process to copy the patterns in these masks to the wafer is called lithography. Chapter 9 discusses the basics of the lithography process together with its current limitations and potential future solutions.

During the past six decades of scaling, the transistor sizes are now so small that their fabrication has become extremely difficult. Advanced chip fabrication plants require investments of more than 20 billion US$ (!!). Chapter 10 discusses the basic process steps in the fabrication of integrated circuits. These steps are then used to build the different layers required to fabricate the transistors and their interconnections. Starting with a description of a conventional five-mask MOS process, the reader is gradually introduced into state-of-the-art fabrication of complex sub-16 nm CMOS (FinFET) chips.

Performance improvements have long been one of the most important driving factors for the continuous development of new integrated circuits. Chapter 11 explains that less power is not only required in the development of battery operated products, but that it is also a must in every other chip application. Next, it describes the impact of scaling of the individual devices (transistors) and interconnections on the overall chip performance and power characteristics. Because smart phones, tablet computers and many other mobile gadgets have become extremely popular, a short introduction on battery technology is included as it has a lot of impact on the standby and operating times of these devices.

Developments in IC technology now facilitate the integration of complete systems on a chip, which currently contain several hundreds of millions to more than eight billion transistors. Therefore, not every chip on a wafer will be fully functional. Failing chips may be caused by random defects during the fabrication process, by critical mask dimensions, by design errors or by limited or too small design margins. The number of functional good dies per wafer, expressed as a percentage of the number of potential good dies per wafer, is called *yield*. Yield increase means cost reduction. Chapter 12 presents an overview of state-of-the-art tests and techniques that support testing. It also describes a simple yield model to explain yield and yield mechanisms.

When the test has identified failing chips on the wafer, it is important to find the cause of that failure. This can be an extremely complex and time-consuming task. A lot of debug and failure analysis techniques exist to quickly localize the failure and identify the failure mechanism to reduce time-to-market. These are discussed in Chap. 13. Once the diagnosis has been made, the chip can be repaired directly by making and breaking techniques. The ability to physically edit an IC (circuit editing) may reduce the number of re-spins and also helps to reduce time-to-market.

When the chip has been tested and found correctly operating, it needs to be packaged to protect it from chemical and mechanical and other environmental influences. A package also serves as an interface to other electronic devices. Chapter 14 discusses the various package alternatives and technologies and gives an insight into their future trends. Essential factors related to chip production are also examined; these factors include quality and reliability.

Finally, the continuous reduction of transistor dimensions associated with successive process generations is the subject of Chap. 15. The first part of this chapter discusses the complexity and problems of the current chip scaling process. It shows that we have almost reached the limits of practically all involved semiconductor disciplines that are needed to develop a chip. This continuation path of scaling is called *more of Moore*, after Gordon Moore, who was one of the founders of Intel. He predicted in 1965 that, due to scaling, the number of transistors on a chip would double every 18 months to 2 years. This has been known as Moore's law since then.

The second part of this chapter discusses how the electronics evolution is expected to continue in this and the next decade. The slowing down of Moore's law is compensated by an increasing combination of micro- and nanoelectronical, electromechanical, chemical and/or microbiological technologies. This will result in the creation of a wide range of new devices, sensors and actuators. These devices will increasingly find their way into all commercial and military markets. In many of the existing applications, such as automotive, communication and Internet of things (IoT), these nanosystems will create new functions and features and/or increase the level of integration for further reduction of size, weight, power and cost.

However, particularly the introduction of these devices in biology with an emphasis on biotechnological and biomedical applications, of which a few existing examples are given in this final chapter, will create an enormous potential of compact electronic devices and systems that will impact human life, with respect to safety, environment, comfort, health and wellness, to beyond today's imagination.

Chapter 5
Digital Circuits and IP

5.1 Digital Circuits

Most integrated circuits are fabricated in a complementary metal oxide semiconductor (CMOS) technology, which features two types of transistors, an nMOS and a pMOS transistor. As the MOS transistor is the basic element of 90% of all integrated circuits, we will repeat some of the material in Part I, because of its importance. Basically, a MOS transistor has three terminals: a gate, a source and a drain (Fig. 5.1). The gate is separated from the channel region by only an extremely thin (t_{ox}) insulating gate oxide.

The transistor operation is based on the amount of current (called I_{ds}) that flows from drain to source. This current is controlled by the voltage on the gate. In fact, this current is defined by the amount of electrons that flow from source to drain. This flow of electrons can only be caused by connecting the transistor terminals to certain voltages. Every *MOS transistor* has a built-in *threshold voltage* (called V_t), below which it does not conduct (no flow of electrons). Only when the gate-to-source voltage (V_{gs}) is larger than the threshold voltage, there will be a conducting channel below the gate oxide isolation that connects the source with the drain. However, as long as the source and drain voltages are equal, there will not be any electron (or current) flow. Next, if the drain is connected to a positive voltage with respect to the source, the *source* will source electrons into the channel, while these are drained from the channel by the *drain*. The larger this drain-to-source voltage (V_{ds}) becomes, the more electrons flow through the channel and the larger the transistor current will be. The width of the transistor channel is denoted by W, while its length is represented by L. It will be clear that a wider transistor channel allows more electrons to flow simultaneously from source to drain resulting in a larger current. However, the longer the transistor channel, the more resistance the electrons face on their path from source to drain and the lower the current will be. We will now express the *transistor current I_{ds}* in relation with the terminal voltages (V_{gs} and V_{ds}), the transistor sizes W and L and the threshold voltage (V_t):

© Springer International Publishing AG, part of Springer Nature 2019
H. Veendrick, *Bits on Chips*, https://doi.org/10.1007/978-3-319-76096-4_5

Fig. 5.1 Basic architecture
of a MOS transistor

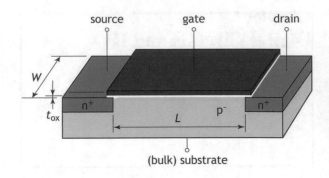

$$I_{ds} = \frac{W}{L} \times \beta_\square \times \left(V_{gs} - V_t - \frac{V_{ds}}{2} \right) \times V_{ds}$$

The threshold voltage depends on the amount of channel dope. *Random doping fluctuations (RDF)* lead to a variability in the threshold voltage of identical transistors. This is particularly an issue in the operation of analog circuits.

$\beta_\square = \mu_n.C_{ox} = \mu_n.k_0.k_{ox}/t_{ox}$ represents the *transistor gain factor*. It is proportional with the *mobility* μ_n of the electrons in the channel, with the *dielectric constant* k_{ox} of the gate oxide, but inversely proportional with the *gate-oxide thickness* t_{ox}.

The speed of a circuit is directly related to the transistor current. Faster circuits require more current. The above current expression shows that one of the ways to increase the current is to increase $\beta_\square$. For many technology generations, this has been achieved by reducing the gate-oxide thickness t_{ox}. This oxide thickness has now become so small (only five to ten atom layers), that it leaks current from the transistor gate to its channel, even when the transistor is switched off (does not conduct). This is one of the most important transistor *leakage currents*, which is particularly a problem in mobile devices where it limits the standby time. The use of *high-k dielectrics* in advanced CMOS processes increases k_{ox}, which allows a larger t_{ox} and reduces the gate-oxide leakage component.

To integrate a maximum number of transistors onto a single chip, most transistors use the minimum possible channel length L defined by the technology. In a 28 nm CMOS technology, the minimum channel length is close to 25 nm. In a 45 nm technology, it is close to 40 nm, etc. Technologies are developed for a fixed supply voltage V_{dd}, which is currently close to 1 V. Transistor terminal voltages V_{gs} and V_{ds} can never be higher than V_{dd}. So, if we want to design a faster circuit in a given technology, we have to increase the transistor current by increasing the transistor width W (see the above expression). A wider transistor has a larger drive capability in digital circuits. In analog circuits an increase in the transistor width may lead to a larger sensitivity of an amplifier, for example. So, optimising the widths of the various transistors that constitute a certain circuit 'gives' this circuit its required performance.

As discussed in Part I, a *digital circuit* only operates on two discrete values: a logic '0' and a logic '1'. All digital circuits, whether they are logic or memory, are

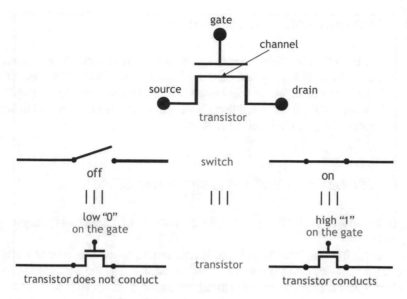

Fig. 5.2 A transistor used as a switch

Fig. 5.3 Example of a logic function: $Z = a$

realised with transistors. In these circuits the transistors are either switched on (conducting, which corresponds to a logic '1') or completely switched off (corresponding to a logic '0'). In digital circuits the transistors are thus only used as switches. Figure 5.2 shows how an nMOS transistor is used as a switch.

This transistor does not conduct ($\equiv$ is off) when its gate voltage is set to low level (logic '0'), below the threshold. However, it does conduct ($\equiv$ is on) when its gate voltage is set to high level (logic '1'), far above its threshold.

Digital circuits operate by using logic functions. Figure 5.3 shows the *logic function* of the so-called *identity gate* or *equality gate*: $Z = a$, where a is the input of the circuit and Z represents its output.

In this example, the output function is equal to the input, because when switch a is off ($a = 0$), the light Z is off ($Z = 0$) and the light is on when the switch is on. This is the most elementary logic function. Most basic logic functions are realised by single logic circuits, which we also call logic gates. These are the subjects of the next subsection.

5.2   Basic Functions and Logic Gates

Like a wall is built from bricks, so is a digital function (multiplier, filter, microprocessor, etc.) built from logic gates. Each *logic gate* is built from transistors which
create its logic function. We will first present a few simple everyday examples of
generally known logic functions, although we are not aware of them, and then find
the logic gates that perform identical functions.

5.2.1   The Inverter (NOT Gate) and Driver Circuit

The first example is to illustrate the logic function of an automatic night light
(Fig. 5.4).

The daylight, generated by the sun, is represented by a. Z represents the night
light. During the day time (a = on; a = 1), light is captured by the sensor, which turns
the night light off (Z = off; Z = 0). During the night, when there is no longer light
captured by the sensor (a = off; a = 0), the night light will be switched on (Z = on;
Z = 1). In other words, the function of the night light is right opposite from the day
light function. When the 'day light is on', the night light is off and vice versa. This
is a so-called inverse function. Mathematically we express this as:

$$Z = \text{NOT } a, \quad \text{or shortly } Z = \bar{a}.$$

The logic gate that performs this logic function is called a *NOT gate* or simply an
inverter. The symbol of an inverter is shown in Fig. 5.5.

A *truth table* represents the logic function in the form of a table. In this particular
example of the inverter, we see that its output Z has the inverse value of its input a:
when a = 0, Z = 1 and vice versa. The figure also shows that the inverter is built from
only two transistors. When input a = 1, the pMOS transistor is off while the nMOS
transistor is on (conducts) and pulls output Z to ground V_{ss} (Z = 0). On the opposite,

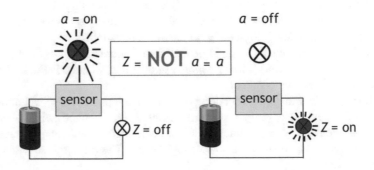

Fig. 5.4 Example of a light sensor representing the logic function of an inverter

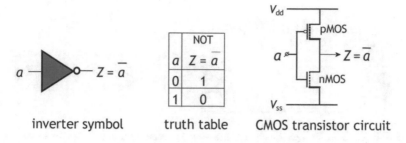

Fig. 5.5 Symbol for an inverter, its truth table and its CMOS realisation

when input $a = 0$, the nMOS transistor is off while the pMOS transistor conducts and pulls output Z to the high V_{dd} level ($Z = 1$). The inverter is the most simple basic logic gate that exists. All other single logic gates are more complex and have the nMOS and pMOS transistors replaced by a combination of serial and/or parallel connected transistors. But every single logic CMOS gate always generates an inverse logic function.

5.2.2 The AND and NAND Gate

Consider a motion-activated night light at the front door. Its logic function will be discussed at the basis of Fig. 5.6.

The sensor includes both a light detector and a motion detector, most commonly an infrared detector. The sensor will only switch on the light, when, first, it is dark, and second, when it detects some movement at the same time. The darkness is represented by a, while motion is represented by b. The light Z will only turn on ($Z = 1$) when there is both darkness (switch a is closed: $a = 1$) **and** motion (switch b is closed: $b = 1$). So we can shortly express this as a logic AND function: $Z = a$ **AND** b or mathematically: $Z = a \cdot b$. An AND function is represented by a dot ($\cdot$).

There exists also a logic function that does right the opposite from the AND function, called a NAND function, where Z becomes low when a and b are high: $z = \overline{a.b}$. $a.b$ means a AND b, but in combination with the bar on top of $a.b$, it means NOT a AND b. The left part of Fig. 5.7 shows the symbols and truth table for the AND and NAND gates. The NAND symbol is identical to the AND but has an additional small circle at its output, which resembles its inverting property. Actually, a *NAND gate* performs the inverse function of an *AND gate*. Both AND and NAND gates in the figure have only two inputs a and b. They are therefore called: two-input AND (AND2) and two-input NAND (NAND2) gates, respectively.

Two inputs allow four different combinations of the 0's and 1's. The truth table shows that the output Z of an AND2 gate only equals 1 when both inputs a **and** b are 1. The output of the NAND2 gate is the inverse of the AND2 gate output: it is 1 for all combinations of 0's and 1's for a and b, except for $a = 1$ and $b = 1$. Because

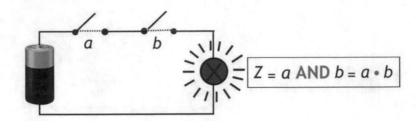

Fig. 5.6 Example of the logic function of a motion-activated night light at the front door

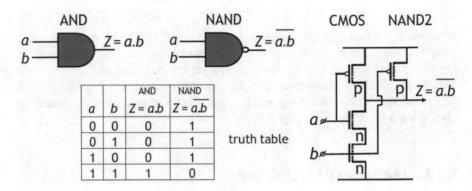

Fig. 5.7 Symbol and truth table for an AND and a NAND gate and CMOS transistor implementation of a two-input NAND gate

every single logic CMOS gate always creates an inverse function, a CMOS implementation of a NAND gate is simpler than that of an AND gate. Therefore only the CMOS circuit of the two-input NAND gate is shown in Fig. 5.7. When we connect its output to an inverter (Fig. 5.5), then both gates together form an AND2 gate.

Next to AND and NAND functions, also OR and NOR functions exist.

5.2.3 The OR and NOR Gate

Consider the example of an elevator in a hotel. We will focus on the logic function Z of its door, which is illustrated in Fig. 5.8.

This door (Z) must open ($Z = 1$) when somebody (a) wants to leave the elevator ($a = 1$) **or** when somebody else (b) wants to enter the elevator ($b = 1$). So, we can express this as a logic OR function: $Z = a$ **OR** b or mathematically: $Z = a + b$. So, an OR function is represented by a plus (+). There exists also a logic function that does right the opposite from the OR function. This is called a NOR function, where Z becomes low when a or b is high: $z = \overline{a + b}$.

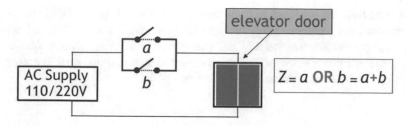

Fig. 5.8 Example of the logic function of an elevator door

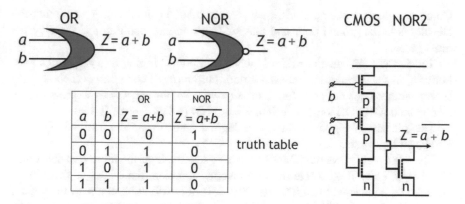

Fig. 5.9 Symbol and truth table for an OR and a NOR gate and CMOS implementation of a two-input NOR gate

The left part of Fig. 5.9 shows the symbols and the truth table of the OR and NOR gates. The NOR symbol is identical to the OR but has an additional small circle at its output, which resembles its inverting property.

Identical to a NAND function being the inverse of an AND function, a *NOR gate* performs the inverse of an *OR gate*. It could therefore also be performed by an OR gate followed by an inverter. Both OR and NOR gates in the figure have only two inputs a and b. They are therefore called: two-input OR (OR2) and two-input NOR (NOR2) gates. The truth table shows that for an OR2 gate, the output Z equals 1 when one of its inputs a **or** b is 1 or when both inputs are 1, which means in three out of the four input combinations. The output of the NOR2 gate is the inverse of the OR2 gate output: it is 0 in three out of the four input combinations and only 1 when both a and b are 0. Again, as every CMOS gate generates an inverse function, only the CMOS implementation of the NOR2 gate is shown in the right part of Fig. 5.9.

Next to the two-input versions of the OR, NOR, AND and NAND gates, there exist versions with more than two inputs, e.g. NOR3, AND4, etc. In the example of an AND4 gate, its output will only be 1 ($Z = 1$) when all four inputs are 1 ($a = 1$ **and** $b = 1$ **and** $c = 1$ **and** $d = 1$): $Z = a \cdot b \cdot c \cdot d$. In case of a 4-I OR gate, its output will only be 1 ($Z = 1$) when only one or more of its inputs are 1 ($a = 1$ **or** $b = 1$ **or** $c = 1$ **or** $d = 1$): $Z = a + b + c + d$.

A logic function needs not be built from (N)OR, (N)AND or INVERT gates only. There also exist combined gates that can be very complex. As shown in Figs. 5.7 and 5.9, both the NAND2 and a NOR2 gate can be built from only four transistors. Complex logic gates, however, may even use more than 30 transistors. Therefore we will first shortly look at the basics of combined gates.

5.2.4   Combined Gates

Consider again the example of the elevator in a hotel (Fig. 5.10). Assume that the elevator is at the ground floor and that we focus on the condition that the elevator must go up.

The elevator (Z) can go up ($Z = 1$) when the door (a) is closed ($a = 1$) **and** somebody (b) in the elevator has pressed a floor button ($b = 1$) **or** when somebody else (c) on another floor has pressed a button ($c = 1$). So, we can shortly express this as a combined AND-OR logic function: $Z = a$ **AND** (b **OR** c) $= a \cdot (b + c)$.

The left part of Fig. 5.11 shows the logic schematic and the truth table of this combined logic gate.

The right part shows the CMOS transistor schematic of the inverse of this combined logic gate. To generate the non-inverted function, the output of this circuit needs to be connected to an inverter (Fig. 5.5). It is still not a complex logic gate, because it only uses eight transistors, including the inverter (not drawn in Fig. 5.11).

A chip may include many different complex digital blocks. We also call them *digital cores*, *logic cores* or *logic blocks*. A multiplier and an adder are relatively small cores, which may each contain several thousand to ten thousand simple and complex logic gates. A microprocessor core, however, may consist of several hundred thousand to more than a few million logic gates, while a complex chip (IC), today, may contain several hundred million logic gates.

In the next paragraph, we will demonstrate how logic gates are used to build an adder and multiplier, respectively.

5.2.5   Drivers (or Buffers)

Most of the signals on a chip only need to travel locally, from one logic gate to another. These short interconnections represent a small load to the logic gates. However, several data signals have to travel relatively long distances inside a core or across the chip. Many designs use local or global buses (see Part I, Sect. 2.6) to transfer data between the various cores on a chip. A *bus* is just a group of wires (*bus lines*) (e.g. 32 for a 32-bit processor) that run from one position on a chip to another. These wires form a much larger capacitive load to the logic gate whose signal must be transported. A normal logic gate is too weak to 'carry' such a heavy load. Therefore, a library has each logic gate available with a wide range of *drive strengths*

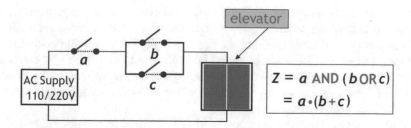

Fig. 5.10 Example of a logic function of the elevator itself

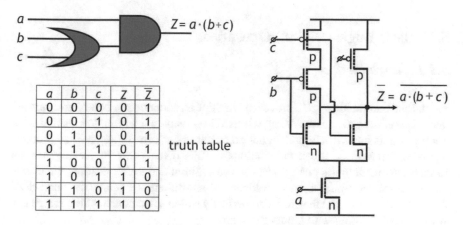

Fig. 5.11 Logic symbols, truth table of a combined logic gate and the CMOS implementation of its inverse function

to support both high-speed and low-power designs. For example, a two-input AND gate may be available in five drive strengths, as a one-time drive (1×drive), a ½×drive, a 2×drive, a 4×drive and an 8×drive. The ½×drive cells are often available for low-power applications that need to run at relatively low frequencies. Preferably the smallest possible drive strength should be used to keep both the area and power consumption low. The larger drive cells should only be used in the critical timing paths.

Today, the libraries also support low-standby power applications, which require the use of library cells with much less leakage currents (leakage power): the higher the threshold voltage V_t, the lower the leakage. Therefore all libraries offer all cells with a standard-V_t transistors as well as with high-V_t transistors.

For signals that need to travel a very long distance, even the 8×drive cells may be too weak, resulting in too slow signal propagation. Therefore libraries also include *driver cells* (also called *drivers* or *buffers*) that do not perform any logic function but are only used to boost the drive strength of the signal to be transported far across the chip. A driver cell is usually built from two cascaded inverters. Also driver cells come with various drive strengths. With the first inverter built from transistors with 'normal' sizes, the second inverter acts as the real driver and has large to very large transistors depending on the required drive strengths. *Clock drivers*, for example, are commonly used to drive the clock signal globally through a digital core to the individual flip-flops. Particularly global bus lines need to be driven by extremely large drivers, which may have 50×drive strengths. *Output buffers*, which need to drive bond pads, board wiring and inputs of other system components, can even have 300×drive strengths.

With this short discussion on drivers, we close this section on basic functions and continue with a discussion how to build larger logic blocks from them.

5.3   Basic Logic (Digital) Operations

5.3.1   Introduction

All the bits that run through a processor, multiplier, adder or any other logic core are generally referred to as *data*. Many ICs used in consumer, communication and computer products include a digital signal processor to perform audio or video processing or to perform complex calculations on this data. Many of these processor functions use arithmetic operations, such as addition and multiplication, which can be performed on digital signals. Adders and multipliers are relatively small logic functions on a chip, and they are also perfectly suited to demonstrate the basics and simplicity of working with binary numbers.

When we make an addition or multiplication, we are used to do so in the decimal number system, e.g. $12 \times 15 = 180$. However, addition and multiplication on a digital chip are performed in the binary number system. In Part I, the conversion from a decimal number into a binary number was already explained. In this section we will explain how to 'add' and 'multiply' in the binary number system. Next, we will build a multiplier to illustrate the use of logic gates to create larger blocks.

5.3.2   Binary Addition and Building an Adder

Assume we want to make the following decimal addition: $6 + 5 = 11$ (on the left in Fig. 5.12) and represent its binary counterpart with 4-bit binary words.

When we add the two decimal digits 6 and 5, it results in a sum digit $S_0 = 1$ in the result and a carry digit $C_0 = 1$ and a sum digit $S_1 = 1$. This is similar for the binary number system. The 4-bit binary representation of 6 is the binary word 0110, while that of 5 is 0101. Every time we add two individual bits (a_i, b_i) of each word, we create a sum bit (S_i) and a carry bit (C_i). When both inputs are 0, the sum bit is 0 ($0 + 0 = 0$). When only one of the two input bits is 1, then the sum bit is 1 ($0 + 1 = 1$, $1 + 0 = 1$). When both inputs are 1, the sum bit is 0 ($1 + 1 = 0$), and the carry bit becomes 1. Figure 5.13 shows the function of this *sum* and the *carry* bit for a

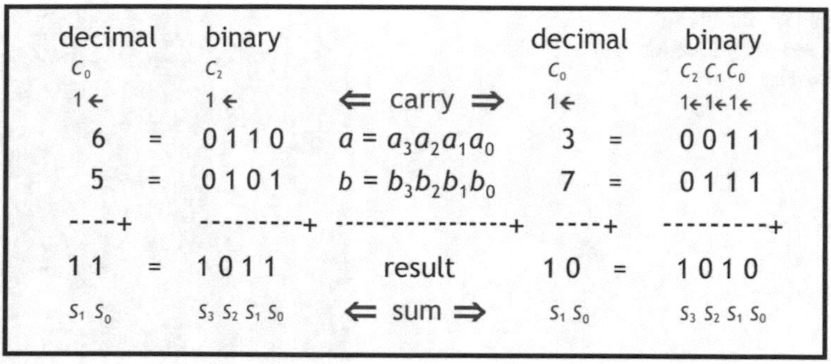

decimal	binary		decimal	binary
C_0	C_2		C_0	$C_2\,C_1\,C_0$
1 ←	1 ←	⇐ carry ⇒	1 ←	1←1←1←
6 =	0 1 1 0	$a = a_3a_2a_1a_0$	3 =	0 0 1 1
5 =	0 1 0 1	$b = b_3b_2b_1b_0$	7 =	0 1 1 1
----+	---------+	-----------------+	----+	---------+
1 1 =	1 0 1 1	result	1 0 =	1 0 1 0
$S_1\,S_0$	$S_3\,S_2\,S_1\,S_0$	⇐ sum ⇒	$S_1\,S_0$	$S_3\,S_2\,S_1\,S_0$

Fig. 5.12 Basic binary addition

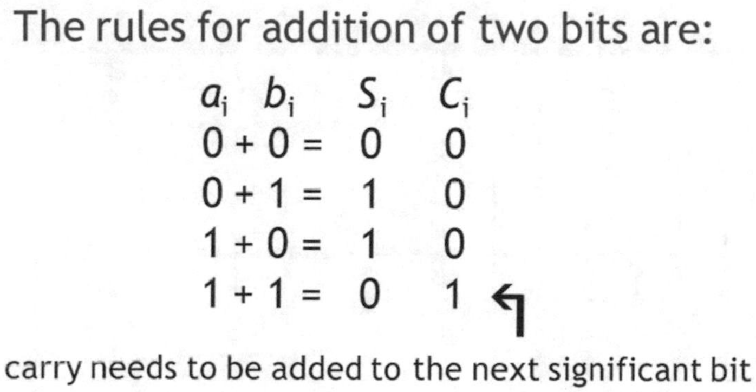

The rules for addition of two bits are:

$$
\begin{array}{cccc}
a_i & b_i & S_i & C_i \\
0 + 0 = & & 0 & 0 \\
0 + 1 = & & 1 & 0 \\
1 + 0 = & & 1 & 0 \\
1 + 1 = & & 0 & 1 \\
\end{array}
$$

carry needs to be added to the next significant bit

Fig. 5.13 Adding two bits a and b creates the sum and a carry function of a half-adder

combination of the two input bits a_i and b_i. This function is called a *half-adder*, because it only adds two bits.

In the example of the binary version of the decimal addition, $3 + 7 = 10$ (on the right in Fig. 5.12), we start adding the least significant (most right) bits a_i and b_i of both binary numbers first, $1 + 1 = 0 (=S_0)$, plus we create an additional 1 $(=C_0)$ that we need to carry one bit position to the left because it has a larger weight. We simply call it the *carry*. For the second least bit position, we need to add: 1 $(C_0) + 1 + 1$, which results in a 1 $(=S_1)$ and another 1 $(=C_1)$. For the third bit position, we now need to add: 1 $(C_1) + 0 + 1$, generating a 0 $(=S_2)$ and a 1 $(=C_2)$. For the most significant bit position, we need to add: 1 $(C_2) + 0 + 0$, generating a 1 $(=S_3)$ and no further carry anymore.

So, when adding two binary numbers, we need to add three bits per bit position: both individual bits a_i and b_i of the binary numbers plus the carry (C_i) from the previous bit position. For this we need a full-adder, instead of a half-adder. A *full-adder* can add three bits and uses the above carry (C_i) as an input, which we now call carry-in c. So it has three inputs a, b and c, and it generates again a sum (S_i) and a

Fig. 5.14 Truth table for
the sum and carry function
of a full-adder

a	b	c	S_i	C_i
0	0	0	0	0
0	0	1	1	0
0	1	0	1	0
0	1	1	0	1
1	0	0	1	0
1	0	1	0	1
1	1	0	0	1
1	1	1	1	1

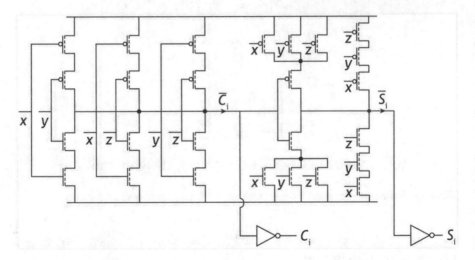

Fig. 5.15 Direct CMOS implementation of the sum and carry-out function in the full-adder gate

carry-out (C_i) signal. Next we can create a truth table for both functions of the full-adder gate as shown in Fig. 5.14.

In this truth table, the sum (S_i) is 1, when one of the three inputs is 1 or when all three inputs are 1. The carry-out (C_i) is 1 when two of the three inputs are 1 or when all three inputs are 1. Figure 5.15 presents the CMOS transistor schematic of the full-adder to show the difference in complexity with, say, the simple 2-I NAND gate of Fig. 5.7. The full-adder contains 30 transistors, while the 2-I NAND gate is built from only 4 transistors.

The symbol of the full-adder is given in Fig. 5.16.

Now we have created the full-adder gate, we can easily build a complete adder function with it. This is explained in the next subsection, in which we focus on the implementation of a binary multiplier, in which an adder is used to perform the final result.

Fig. 5.16 Logic symbol of
a full-adder gate

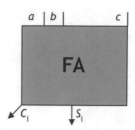

5.3.3   Building a Multiplier

A *multiplier* needs to operate on two numbers. The first number, the *multiplicand*, has to be multiplied by the second number, the *multiplier*. In a binary multiplier, the individual bits of the two operands need to be multiplied, and the results need to be added. Therefore the larger part of a multiplier consists of an *adder*.

Assume we want to implement the following decimal multiplication: $16 \times 13 = 208$. We start with the decimal to binary conversion of each of the two operands (see Fig. 5.17).

Multiplication in the binary system is easy. We need to multiply the multiplicand by the multiplier bit by bit. When the bit in the multiplier operand is a 0, we only need to shift the multiplicand one bit position to the left but need not add it. When the next bit in the multiplier is a 1, we need to shift the multiplicand again one bit position to the left and add the shifted multiplicand to the previous result. This process is repeated until we have arrived at the most significant bit of the multiplier operand. A binary multiplication therefore only consists of simple shift-and-add procedures, which is clearly shown in the figure. The multiplication of two individual bits of the multiplicand and the multiplier is easy, and the result is called the product. Figure 5.18 shows the rules for the product P of two individual bits.

From the table, we can see that the product P is only 1 when both inputs a **and** b are 1. This is equal to a logic AND function which can be represented by an AND gate (see Fig. 5.7). So, looking back at Fig. 5.17, a binary multiplication consists only of many cross products of the individual bits of the multiplicand and the multiplier operands, which need to be added to perform the final addition to create the final product.

Figure 5.19 shows the architecture of a basic 3 by 3 bit multiplier, for educational purposes. On the cross roads of each individual a_i and b_j signal line, there is an AND gate to perform their product. This results in an array of AND gates. Next, all outputs of these AND gates need to be added in an array of full-adders, to perform the addition to create the final product. In this small 3 by 3 bit multiplier, the full-adders do not actually form an array. However, in a 32 by 32 bit multiplier, this involves a large array of about 512 full-adders.

It takes a certain amount of time before the final product is 'ready', simply because the various different signals have to propagate through different *logic paths*. These are also called *data paths* or *delay paths*, because each path represents a

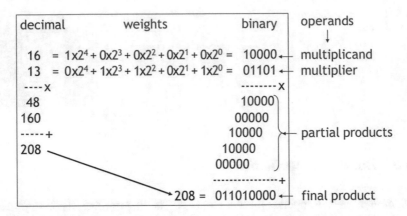

Fig. 5.17 Binary representation of a multiplication

Fig. 5.18 Multiplication of two bits

The rules for the product of two bits are:

$$a \quad b \quad \text{product } (P)$$
$$0 \times 0 = 0$$
$$0 \times 1 = 0$$
$$1 \times 0 = 0$$
$$1 \times 1 = 1$$

$P = 1$, when $a = 1$ **AND** $b = 1$, so $P = a \cdot b$

certain delay that a signal faces when propagating through it. The signal in the blue delay path in the figure only needs to propagate through an AND gate and one full-adder (FA) before it arrives at the bit position p_1 of the final product. The final product is only 'ready', when the signal through the *worst-case delay path* has arrived at the *most significant bit* in the final product. In this example, this is bit p_5. The purple delay path shows a worst-case delay path example, in which the data needs to propagate through the AND gate and four full-adders.

Today's signal- and microprocessors require much more accuracy, and their digital signals are represented by 16-, 32- or even 64-bit words. Multiplication of two 64-bit words results in an array of 64 by 64 AND gates. The outputs of all these AND gates must then be added in a huge array of full-adders to generate the final product. Such multipliers therefore show relatively long worst-case delay paths, meaning that the execution of complex multiplications on a chip will take some time.

Next, the result of the multiplication, the final product, needs to be stored in a memory or needs to be sent to the output of the chip. We therefore need to give the multiplier sufficient time to complete its final product; otherwise the incomplete (wrong) result will be stored our send to the output. This requires a certain timing control, not only of the multiplier core but also of all other logic and memory cores on the chip. Signals within and between these cores need to be synchronised, in order to transfer these signals at the right time to the right input, core or output.

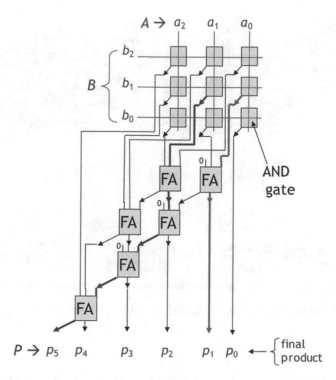

Fig. 5.19 Example of a 3 by 3 bit multiplier $P = A * B$

5.4 Synchronous Designs and Flip-Flops

In the previous section, it has been explained that a logic circuit may contain many different logic gates: INVERTER, 2-I AND, 2-I OR, 2-I NAND, 2-I NOR, 3-I NAND, 3-I NOR, HALF-ADDER, FULL-ADDER, etc. These logic gates are most commonly used to perform certain logic and digital operations: add, multiply, compare, digital filter, digital (signal) processor, etc. Most digital integrated circuits are *synchronous designs* in which the communication within and between the logic cores is controlled by a clock signal, by using latches or flip-flops. A *latch* often contains two inverters with positive feedback, such that once they are forced into a logic state, its positive feedback loop will keep it there until it is forced into another state. A *flip-flop* is usually built from two latches, which can temporarily hold the data for one clock period. Figure 5.20 shows an example of a flip-flop. It consists of a *master* and a *slave*. Pass gates $pass_1$ and $pass_2$ function complementary, meaning that when one of them is on (passes), the other one is off (isolates) and vice versa. When the clock signal ϕ is low, $pass_1$ conducts and $pass_2$ is off and the positive feedback loop in the master that is formed by the two inverters and $pass_2$ is

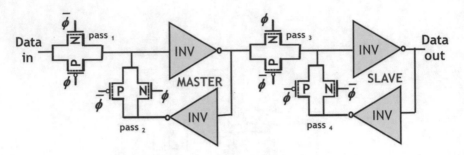

Fig. 5.20 Example of a flip-flop in CMOS technology

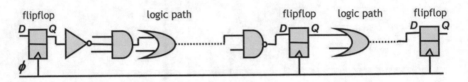

Fig. 5.21 Example of the use of flip-flops within a synchronous logic core (flip-flop)

interrupted when sampling the input signal (data in) into the master. Once the data has been sampled into the master, the clock signal ϕ goes high, thereby switching off pass$_1$ and switching on pass$_2$. Now the master acts as a latch, which will maintain, due to the positive feedback loop, the state that had just been sampled from its input.

Next, a similar sequence of operations by pass$_3$ and pass$_4$ will copy the output of the master into the slave. So, after one full cycle of the clock signal ϕ, the flip-flop input signal has arrived at its output.

To support a safe operation of the logic functions and guarantee a correct signal flow through the logic gates and logic cores, flip-flops are used to synchronise the data flow through and in between these cores. This synchronisation is controlled by the clock signal. So, in a synchronous design, this clock signal controls the data flow to the chip, through the chip and from the chip, like traffic lights control the traffic in a city. Figure 5.21 demonstrates how these flip-flops are used within a core to cascade individual logic paths. The flip-flops temporarily store the data and let it go on clock demand. At any time the positions and values of all data samples in a synchronous digital block and chip are known (by simulations).

Although the clock network and flip-flops in a synchronous design do not contribute to the IC's functionality (they are only there to control the signal flow), they can take a relatively large part of both the area and power consumption. In certain high-performance processor designs, the clock network, including the flip-flops, may consume close to two third of the total chip power!

5.5   Digital Core and Chip Design

In the previous chapter, we have explained that a chip is built from various different cores. Some of these cores are *re-usable cores*, which may be used on various chips in a variety of applications. These are often referred to as *intellectual property* (*IP*) cores. They may be available from an in-house library or from an external supplier, because they represent a frequently used functionality. Other cores need to be designed from scratch by using the existing logic gates, such as explained in the previous paragraphs. These logic gates are implemented as standard cells (see also Part I, Sect. 2.6.1) that are picked from a library by so-called synthesis tools. Today these libraries may contain a large variety of cells:

- More than 1500 standard cells, like NAND, NOR, full-adder, flip-flops and drivers with various drive strengths and threshold voltages, as discussed in Sects. 5.1, 5.2 and 5.3
- Several hundred IP cores, e.g. analog cores, multipliers, adders, signal processors, controllers, interfaces, etc.
- Different types of memory cores, e.g. random access memories (SRAM, DRAM), non-volatile memories (ROM, E(E)PROM, flash, etc.)
- 50 or more I/O cells (inputs, outputs, different drive strengths, compliant with the various standard interfaces, e.g. USB, PCI, DDR, and compatible with the different voltages)
- All the characteristics of these cells, e.g. function, speed, delay, power consumption, etc.

As stated before, all logic cores are built from *standard cells*. Each standard cell represents a logic gate, which realises a certain logic function, a flip-flop or a driver circuit to drive long signal lines with heavy load capacitances. They are called standard cells, because all cells, even though they have different drive strengths, have the same height with their supply and ground rails exactly at the same position (Fig. 5.22). After they are placed and connected by a so-called place-and-route tool, these cells form continuous rows, with the supply and ground lines automatically connected.

Most place-and-route tools require that the width and height of the library cells are equal to an integer number of second metal routing tracks that fit across the cells. The cells in Fig. 5.22 correspond to a 9-track library cell height.

Figure 5.23 shows the standard-cell layout of a *flip-flop* (see Sect. 2.1 in Part I). Almost all systems, today, are synchronous, which means that all electronic circuits of the system are controlled by one or more clock signals. These clock signals are fed to flip-flops, which hold the data and let it go on clock demand (e.g. when clock switches to high).

The different colours in the layouts usually correspond with the different layers from which these circuits are built in the CMOS process. Figure 5.24 shows only three of these layers, just to explain the relation between the layout and these layers. The various layers that are used to fabricate a complete chip are isolated from each

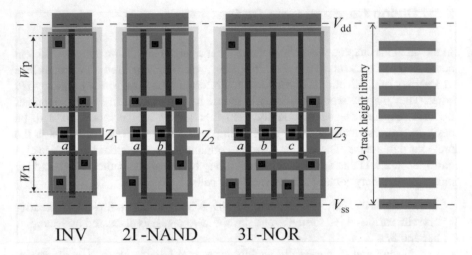

Fig. 5.22 Example of four different standard cells from a digital library

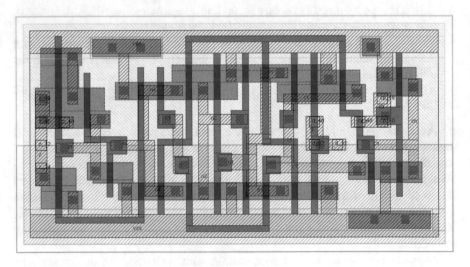

Fig. 5.23 Example of a standard-cell layout of a flip-flip. (Photo: Cogenda)

other by so-called oxide isolation layers. These layers are not included in the figure since they would block the visibility of the layers below them.

Figure 5.25 shows the design flow used to design a digital core. First, its function is specified and modelled in a high-level language (VHDL, Verilog or C code).

At this level, only the function of the core is described. It does not say anything about how the core is implemented. This description (code) can be read by a *logic simulator* to verify the functionality. Extensive simulations are needed, because the verification may require many iterations (code corrections and re-simulations) before a correct operation of the core can be guaranteed. Next, the final code is read

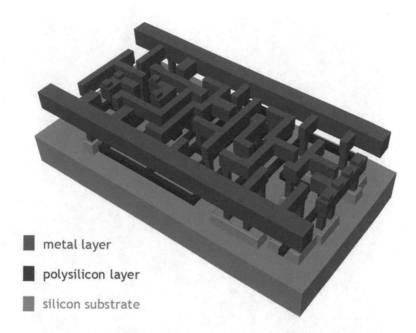

metal layer

polysilicon layer

silicon substrate

Fig. 5.24 Geometric representation of several CMOS process layers from which the flip-flop of Fig. 5.23 is built. (Photo: Cogenda)

by a logic *synthesis tool*. This tool is familiar with the function and architecture of many basic building blocks, such that it can translate a relatively short high-level functional description into a *net list*. This net list not only contains a list of all standard cells that are needed to perform the required functionality, it also specifies to which nets the inputs and outputs of all cells are connected. Advanced synthesis tools can also automatically extract larger blocks out of the code, such as a multiplier, a counter or a memory. The result of the synthesis is very much dependent on the designer's specification (code). The implementation could be optimised for lowest power, smallest area or highest performance. Once the net list is generated, the physical implementation must be realised. This is done by a so-called *place-and-route tool*, which, as the name implies, does it in two steps. First the individual standard cells are placed in an array of rows, and then all cells have to be connected according to the net list. During placement of an individual cell, the tool already keeps track with the position of the cells to which it must be connected.

When the placement is completed, the layout is checked with respect to the layout design rules, and its operation and timing (speed) are verified against the functional and specification requirements, respectively. When these requirements are not met, the design description needs to be corrected and resynthesised, etc. So, a core design may consist of several of such iterations before it is operating correctly. The final result is a digital core (IP core; logic core). Figure 5.26 shows an example of a relatively small logic core, in which the rows of standard cells are clearly visible. Specialists in the field of IC design would immediately recognise that this logic core

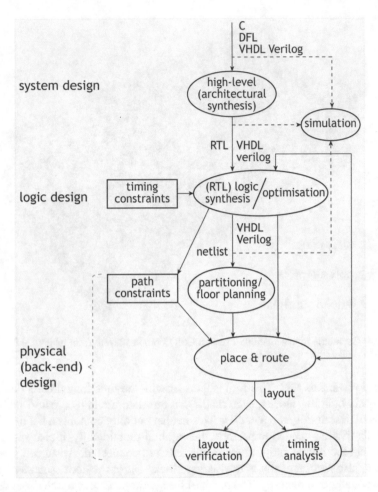

Fig. 5.25 Example of a standard-cell design flow

is implemented in a conventional CMOS technology. The density of the cells in state-of-the-art CMOS technologies would be at least an order of magnitude larger, and seven or more metal layers would be used to connect all these cells. On a photograph of such an advanced standard-cell block, the individual cells and rows would no longer be visible.

5.5.1   Example of Synthesis from VHDL Description to Layout

This paragraph discusses the design steps of a digital potentiometer, starting at the RTL description level (in VHDL) and ending in a standard-cell layout. Figure 5.27 shows the RTL-*VHDL description* of this potentiometer. An explanation of such a

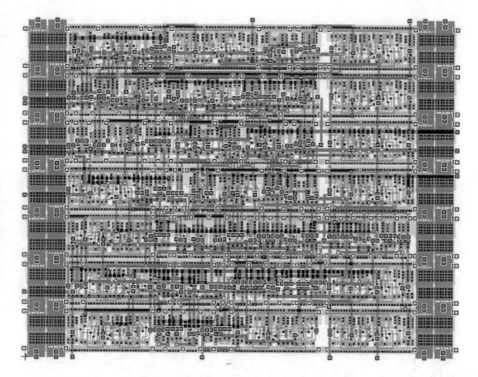

Fig. 5.26 A relatively small digital core, in which the standard-cell rows are clearly visible. (Photo: Mississippi State University)

VHDL description is beyond the scope of this book. Normally, a VHDL course would almost take a week and a month play to become a little experience with it.

Figure 5.28 shows a part of the net list of library cells onto which the potentiometer function has been mapped. A *net list* may contain instances, modules, pins and nets. An instance is the materialisation of a library cell or a module. A module, itself, is built from several instances and their connections. Pins, also called ports or terminals, represent the connection points to an instance or module, and finally, a net represents a connection between pins. The figure shows the different library cells and the nodes to which their inputs and outputs are connected. The underlined nodes with the same colour, in this net list, are connected to the same nets.

The next step is to create the layout of this block. The *place-and-route (P&R) tool* places the net list cells in rows and also creates the interconnections between the pins of the cells (or modules). Due to the growing complexity of IP cores in combination with the need to accommodate higher chip performance, the physical design of these cores becomes a real challenge. To achieve timing closure on such complex blocks with very tight area, timing and power constraints are a difficult task. Cell placement is a critical part of the back-end design flow, as it has severe impact on core area, wire length, timing and power requirements. P&R tools, today, allow area-driven, wire length-driven and timing-driven as well as power-driven

```
LIBRARY IEEE;

USE IEEE.std_logic_1164.ALL;
USE IEEE.std_logic_arith.ALL;
USE IEEE.std_logic_unsigned.ALL;

ENTITY potmeter IS

    GENERIC (par_width: natural := 4;
             operand_width : natural := 12);

    PORT (A, B: IN std_logic_vector(operand_width-1 DOWNTO 0);
          K: IN std_logic_vector(par_width-1 DOWNTO 0);
          Z: OUT std_logic_vector(par_width+operand_width-1 DOWNTO 0));

END potmeter;

ARCHITECTURE behaviour OF potmeter IS

BEGIN

    PROCESS (A, B, K)

        CONSTANT K_max: integer := 2**par_width-1;
        VARIABLE K_int: integer;

    BEGIN

        K_int := conv_integer(K);
        Z <= K*A + conv_std_logic_vector(K_max-K_int, par_width) * B;

    END PROCESS;

END behaviour;
```

Fig. 5.27 RTL-VHDL description of potentiometer

placement and thus allow placement optimisation for various different application domains. A timing-driven placement, for example, can assign higher weights to critical nets to reduce their wire length as well as select faster cells, e.g. with higher drive capability and/or reduced threshold voltage, to reduce the critical path delay, e.g. for high-speed circuits.

The place-and-route tool places the cells of the net list in rows of standard cells, which results in a standard-cell core design of the potentiometer (see Fig. 5.29). This net list and layout are the result of the chosen description of the potentiometer's functionality according to:

$$Z = k \times A + (1-k) \times B$$

This implementation requires two adders and two multipliers. However, an obvious optimisation of the same function may lead to a more efficient implementation. The following description

$$Z = k \times (A-B) + B$$

```
module potmeter_DW01_add_14_1 ( A, B, CI, SUM, CO );
input   [13:0] A;
input   [13:0] B;
output  [13:0] SUM;
input   CI;
output  CO;
    wire n52, n53, n54, n55, n56, n57, n58, n59, n60, n61, n62, n63, n64, n65,
         n66, n67, n68, n69, n70, n71, n72, n73, n74, n75, n76, n77, n78, n79,
         n80, n81, n82, n83, n84, n85, n86, n87, n88, n89, n90, n91, n92, n93,
         n94, n95, n96, n97, n98, n99, n100, n101, n102, n103, n104, n105;
    BF1T1 U5 ( .Z(SUM[2]), .A(A[2]) );
    BF1T1 U6 ( .Z(SUM[0]), .A(A[0]) );
    BF1T1 U7 ( .Z(SUM[1]), .A(A[1]) );
    AO6 U8 ( .Z(n52), .A(n53), .B(n54), .C(n55) );
    AO6 U9 ( .Z(SUM[3]), .A(n56), .B(n57), .C(n58) );
    AO32 U10 ( .Z(n59), .A(n60), .B(n61), .C(n62), .D(n63) );
    AO32 U11 ( .Z(n64), .A(n59), .B(n65), .C(n54), .D(n55) );
    NR2 U12 ( .Z(n66), .A(n67), .B(n68) );
    AO6 U13 ( .Z(n69), .A(A[7]), .B(B[7]), .C(n70) );
    NR2 U14 ( .Z(n71), .A(n65), .B(n63) );
    NR2 U15 ( .Z(n72), .A(n73), .B(n74) );
    AN2 U16 ( .Z(n75), .A(n76), .B(n77) );
    EO U17 ( .Z(SUM[9]), .A(n66), .B(n78) );
    EO U18 ( .Z(SUM[8]), .A(n79), .B(n80) );
    EO U19 ( .Z(SUM[6]), .A(n81), .B(n82) );
    EO U20 ( .Z(SUM[5]), .A(n71), .B(n83) );
    MUX21N U21 ( .Z(SUM[13]), .A(B[13]), .B(n84), .S(n85) );
    IV U69 ( .Z(n84), .A(B[13]) );
    IV U70 ( .Z(n105), .A(A[10]) );
    IV U71 ( .Z(n96), .A(B[7]) );
    IV U72 ( .Z(n79), .A(n95) );
endmodule
```

Fig. 5.28 Part of the potentiometer net list after synthesis with 14 ns timing constraints

requires only two adders and one multiplier. This example shows that the decision taken at one hierarchy level can have severe consequences for the efficiency of the final silicon realisation in terms of area, speed and power consumption.

Although the synthesis process uses tools which automatically generate a next level of description, this process is controlled by the designer. An excellent design is the result of the combination of an excellent tool and a designer with excellent skills in both control of the tools and knowledge of IC design.

Figure 5.25 showed a few more tools than discussed here, but a description is beyond the scope of this book. In Sect. 7.4 we will return to the subject of full-chip design.

The above-discussed standard-cell design approach is not just the only design style for creating integrated circuits. This design style is only economically feasible when the expected market volumes are relatively high. However, certain applications only require a limited number of products. For instance, when a TV producer decides to put a few thousand new TV sets on the market for evaluation of the new features and customer reviews, it does not want to go through the time-consuming complete design trajectory. The following subsection presents a few alternative design approaches to quickly implement such prototype ICs.

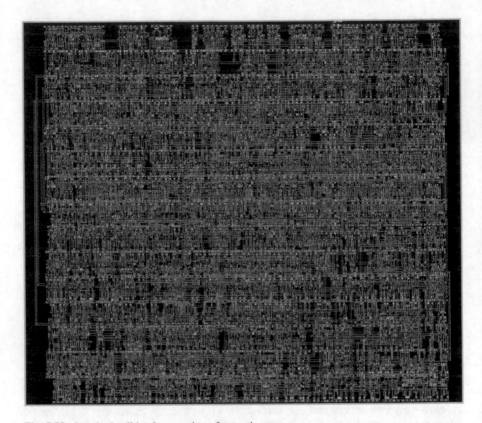

Fig. 5.29 Standard-cell implementation of potentiometer

5.6   Gate Arrays and Programmable Logic Devices

Certain application areas require fast system development with fast turnaround times. These applications cannot live with the long development trajectory of an IC using the above-described standard-cell design flow. There are several solutions for these applications.

The first one is the *mask-programmable gate array* or shortly *gate array* [2]. Most of the fabrication steps of these chips have already been performed, and they usually contain already all transistors and supply lines. These gate arrays are available off the shelf and come with different complexities, e.g. number of transistors and I/Os and the memory capacity. Most commonly, a customer specifies the chip's functionality using the gate array vendor's design tools. Next, the gate array with the required complexity is chosen. The functionality will then be implemented by only defining the patterns in the few final masks. The preprocessed wafers are then again fed into the clean room to fabricate these final contact and metal layers. This will reduce the total processing time to only a few weeks, compared to 7–10 weeks for a normal IC. The popularity of these (mask-programmable) gate arrays reached

Fig. 5.30 General
representation of an FPGA
architecture

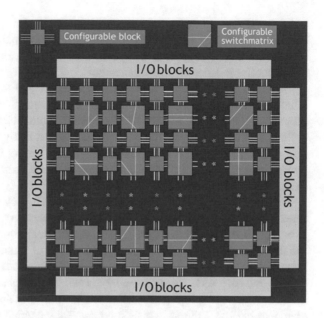

a maximum during the 1990s. The last decade showed a dramatic reduction in new gate array design starts, mainly due to the rapid cost reduction and gate complexity increase of the field-programmable gate arrays. These FPGAs have now almost completely taken the MPGA market.

Another solution to quickly implement an electronic system is to use a *programmable logic device (PLD)*, today, also often referred to as *field-programmable gate array (FPGA)*. These also come in different complexities and may contain a variety of processor and/or memory cores. Initially, FPGAs were used to integrate the *glue logic* in a system. Glue logic is some additional custom logic that connects a number of large blocks or chips together. However, the rapid increase in their complexity and flexibility make them potential candidates for the integration of high-performance, high-density (sub)systems, previously implemented in mask-programmable gate arrays. The potentials of an FPGA will be discussed on the basis of a generic FPGA architecture (Fig. 5.30).

It consists of an array of configurable logic blocks and configurable switch matrices. In this example, each logic block contains a programmable four-input logic gate and includes a small 16-bit memory. The content of this memory determines the function of the logic gate. Each switch matrix contains a set of transistors. Some transistors are connected between horizontal wires, some between vertical wires and some between horizontal and vertical wires. When switched on, they can guide the signals through the chip, in horizontal, vertical or diagonal direction, respectively. Each switch matrix also contains memory cells that are connected to these transistors. A logic one in the memory means that the connected transistor is switched on thereby transferring the data. In other words, both the functionality and the interconnections are controlled by the contents of their respective memory cells.

Today, these architectures consist of a large array of millions of programmable (*re*)*configurable logic block*s (logic cells) and switch matrix blocks. Many FPGAs contain short wire segments for local interconnections as well as long wire segments for 'long distance' interconnections.

Most FPGAs use SRAM memories (next chapter) to store the configuration bits, although there exists also a few who store them in a non-volatile EEPROM or flash memory. All FPGAs that use SRAM for configuration storage need a shadow non-volatile backup memory on the board to be able to quickly download the application into the on-chip configuration memories. Downloading from a software program would lead to relatively large configuration times, whenever the application is started again after power-down.

Next to the configurable logic and switch matrix blocks, many FPGA architectures include dedicated IP cores, digital signal processors (DSPs) [3], microprocessors such as ARM and PowerPC, single and/or dual port SRAMs and multipliers. Finally most of their I/O blocks support a variety of standard and high-speed interfaces. Of course also several dedicated memory interfaces, such as DDR, DDR-2, DDR-3 and SDRAM, are supported. I/Os and interfaces are discussed in Chap. 3.

The above description presented only a flavour of the potentials of current state-of-the-art FPGAs. FPGAs may be used as system emulators or hardware simulators to speed up the development process of complex systems and chips. In this case the system is mapped onto one or more FPGAs, which can then mimic the application to verify its correctness. Next, the correct system can be implemented on a chip using the normal standard-cell design flow.

5.7   Conclusion

An electronic system may consist of a large variety of functions: analog, digital, RF, memories, interfaces, etc. A complex system may require several individual chips, while the functionality of a simple system may be realised on one single chip. Digital circuits usually occupy the largest part of a system or a chip.

This chapter has provided a brief overview on the basics of digital circuits, starting with an explanation on the development of basic logic gates and ending with a description on the development of digital IP. To limit the design time of the digital cores on a chip, their design is supported by a library of standard cells and a lot of tools. Designing a new chip from scratch would take an enormous amount of design time and costs. It is therefore important that the designer's efficiency is continuously improved. This is achieved by creating a lot of IP cores that are part of the library and which can be reused in many different designs.

Next, chip design is supported by a design flow from system-level specification through standard-cell implementation and simulation to chip verification, which consists of a large series of complex tools. This has not prevented that in many complex systems, the design costs per chip have exceeded the fabrication costs. Costs issues are further discussed in Chap. 7.

In most applications, the results of a digital operation must be stored in a memory. Small memories are usually embedded on the chip, while large memories are implemented as stand-alone devices on separate memory chips. Memories have become very popular products in many applications and are an important part of the total semiconductor market, today. The following chapter is therefore devoted to the basic organisation and operation of the most important memories.

Chapter 6
Memory Circuits and IP

6.1 Introduction

Memories are circuits designed for the storage of digital values. In a computer system, memories are used in a large variety of storage applications, depending on memory capacity, cost and speed. Figure 6.1 shows the use of memory storage at different hierarchy levels in a computer system.

The high-level memories directly communicate with the computer's processor. These registers must deal with its high-data communication bandwidth and therefore need high performance, but they are expensive per storage bit. As we move down the hierarchy, both the memory capacity and the access time increase, resulting in a reduction of the cost per bit. *Registers* consist of a small set of memory locations that are part of the central processing unit (*CPU*), often called the *processor*. A *cache memory* acts as an intermediate storage between these ultra-fast processor registers and the main memory and stores the most frequently and/or most recently used data and instructions for fast access.

The computer can only process data that is in the cache and in the main memory. To execute a program, it needs to be copied into one of these memories. All data that need to be processed must be copied in these memories first. The capacities of these memories are therefore very critical to the overall performance of the computer, because they determine how many programs can be executed simultaneously and how much data will be available in time. Advanced microprocessors embed level 1, level 2 and level 3 caches on the microprocessor chip itself to enable the highest performance. The larger these embedded memories, the lower the number of required accesses to the main memories, which have limited communication bandwidth.

In the following text, we will first explain some typically used memory terminology.

There are a few performance parameters that impact memory application: memory capacity, access time and cycle time. *Memory capacity* is usually referred to as the maximum number of bits or bytes that a memory can contain. The *access time*

© Springer International Publishing AG, part of Springer Nature 2019
H. Veendrick, *Bits on Chips*, https://doi.org/10.1007/978-3-319-76096-4_6

Fig. 6.1 Memory
hierarchy in a computer
system

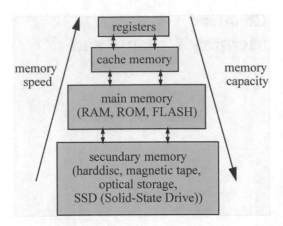

of a memory is the time interval between the moment that the read command is given and the moment at which the data is available at the output pins of the memory. The *cycle time* of a memory is the minimum possible time between two successive accesses. The cycle time is the most important one, since it determines how fast we can read or write data in a sequence from or to the memory.

A memory may constitute a single IC or it may be part of a larger IC. These types are referred to as *stand-alone memory* and *embedded memory*, respectively. The data in a memory is stored as individual bits in the memory cells. These cells are arranged in arrays containing rows and columns, which afford an optimum layout.

There are several different memory types, each with a specific storage method and storage capacity, which depend on the intended application. Memories that lose their data when power is removed are referred to as *volatile memories*. Memories that retain their data in the absence of power are called *non-volatile memories*. The *data retention time* of a memory is the period for which it keeps its data when the supply voltage is removed.

A *random-access memory* (*RAM*) is a volatile memory where the data can be written into (stored) or read from random locations in the memory at (very) high frequencies (usually up to several hundreds of megahertz). We distinguish two categories of RAMs. A *static RAM* or *SRAM* is a volatile memory in which the cell data is maintained by two positive feedback inverters, identical to the operation of a latch (Chap. 5). A *dynamic RAM* (*DRAM*) is also a volatile memory, but now the cell data is stored as charge on a capacitor. Gradual leakage of this charge necessitates periodic refreshing of the stored data (charge) in a DRAM. A DRAM that internally refreshes its own data automatically is sometimes also called a pseudo-static or virtually static RAM.

A *read-only memory* (*ROM*) is a non-volatile memory in which the stored data has been fixed in the memory cells during fabrication and cannot be changed. As a consequence, the data can only be read, which refers to its name. Non-volatile memories further include programmable ROMs, including erasable programmable read-only memories (EPROMs), electrically erasable programmable read-only memories (EEPROMs) and flash memories. Over the last decade, flash memories have gained

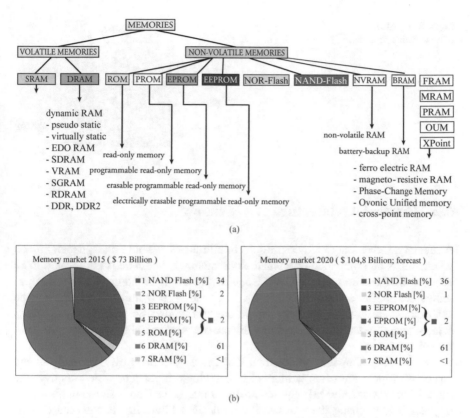

(a)

(b)

Fig. 6.2 (**a**) Overview of different types of memories. (**b**) Relative memory market shares in 2015 and 2020. (Source: IC Insights)

enormous popularity as non-volatile memories. These are often implemented as *memory cards*, e.g. smart media (SM) cards, secure digital (SD) cards and compact flash cards (CF), as (USB) memory sticks and as *solid-state drives* (*SSD*). All non-volatile memories must be able to maintain their data for at least 10 years.

Figure 6.2 presents an overview of the various memories. This figure also shows the respective market shares in 2006 and 2011. As a result of the recent financial crisis, the total memory market of 100 B$ predicted for 2011 seems a little too optimistic.

The large market share gained by the dynamic RAMs (DRAMs) is mainly the result of new high-speed architectures, which make them particularly suited for the growing high-bandwidth memory applications such as games, video and graphics applications, digital cameras, printers, etc. Also the increasing popularity of the NAND flash memories is clearly visible.

The quantification of bits and bytes in this chapter is according to international conventions: 1 Gb equals 1 gigabit (1 billion bits) and 1 Mb = 1 megabit (1 million bits), etc., 1 GB equals 1 gigabyte and 1 MB = 1 megabyte, etc. The word *byte* is a short notation for 'by eight', and so it contains 8 bits.

Fig. 6.3 General
representation of a memory

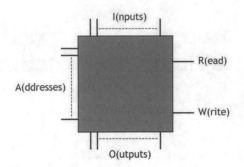

6.2   General Architecture of a Memory

A memory, today, may include millions to billions of memory cells, depending on whether it is an embedded or a stand-alone memory. These cells are located in an array of rows and columns, with each cell having a unique 'address' to enable individual access. To address a cell, the corresponding row and column have to be selected. Therefore the memory arrays are surrounded by address logic, which selects the individual row and column. Figure 6.3 is a general schematic representation of a memory.

The memory shown is obviously a RAM or a flash memory. The read (R) and write (W) inputs are often combined in one single input pin, which controls the mode of operation. A ROM requires no data inputs as its data is stored in one of the physical layers during the process; it only needs address inputs and data outputs during reading. The 'black box' of a ROM is therefore obtained when the inputs (I) and write command signal (W) in Fig. 6.3 are removed.

Various types of memories are discussed in this chapter. Their operation and properties are explained, and potential applications are given. For all memories, every possible combination of address inputs can be decoded. A memory with n address inputs can select 2^n individual addresses, e.g. a RAM with 20 address inputs can select at least 2^{20} (=1 billion) memory cells or words. In the following section, we will discuss a so-called by-1 memory (4 kb × 1), meaning that with each address selection, only one cell (one bit) is accessed. This has been done for educational purposes.

Alternatively, memory cells (bits) may be grouped. In this case, each group or *word* has a specific address. The capacity of a RAM or ROM that is divided into words is specified by the number of words and the number of bits per word. Examples are 1 Gb × 4, 512 Mb × 8, 256 Mb × 16, 128 Mb × 32 and 64 Mb × 64. These five specifications all refer to a 4 Gb memory, which can store over 32,000 newspaper pages or 9 h of MP3 music. Figure 6.4 shows the physical representation of a by-8 memory architecture, where eight memory blocks can be read or written in parallel.

The more parallel accesses a memory allows, the higher the communication bandwidth to interfacing circuits such as signal processors and microprocessors.

Fig. 6.4 Physical representation of a by-8 RAM architecture

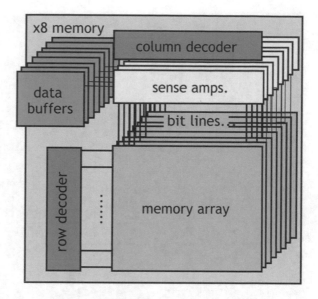

Memories are used to store and read data. When both need to be done at high frequencies, then a random-access memory (RAM) is required. Some applications, however, require memories in which the contents change only once and a while but need high-speed read operations. An example is an MP3 player, in which the music may be stored for a month or a year, but it may be played every day or even several times a day. Flash memories are better suited to support such applications. Because of their non-volatility, they can maintain the music for at least 10 years.

Every memory is implemented as one or more arrays of memory cells, each of which must be accessible for writing data into or reading data from. Therefore, each individual cell has its own address. We will explain the operation of a memory by using the example of a 4 kb × 1 memory of Fig. 6.5.

Its 4096 (=4 kb) cells are organised in an array of 64 rows and 64 columns. There is a memory cell on every crossroad of a row and column. To select one cell, we have to determine in which row and in which column that cell is positioned. Next, we need to select that row and column. To select a single row out of the 64 rows in the example, we need a row address that contains 6 bits ($A_5A_4A_3A_2A_1A_0$), because with 6 bits, we can represent 64 different combinations of ones and zeros. When $A_5A_4A_3A_2A_1A_0 = 000000$, this is decoded by the *row decoder*, which then selects word line x_0, because the binary number 000000 is equivalent to the decimal number 0 (see Part I, Sect. 3.4). When $A_5A_4A_3A_2A_1A_0 = 111111$, the row decoder selects word line x_{63}, because the binary number 111111 is equivalent to the decimal number 63. The same holds for the selection of the appropriate column, which has $A_{11}A_{10}A_9A_8A_7A_6$ as column address bits. Assume that $A_{11}A_{10}A_9A_8A_7A_6 = 0000101$, then the column selector y_5 is activated, connecting bit lines b_5 to the data bus lines *db*.

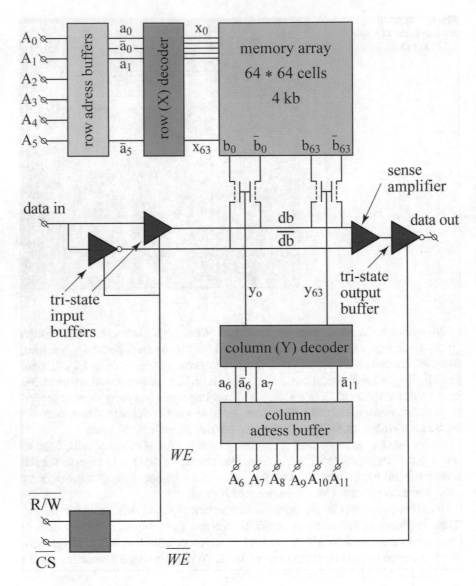

Fig. 6.5 Block diagram of a 4 kb RAM array

When the memory is being written, first the memory itself must be selected through the chip select (CS) signal. This is needed because memories are often part of a *memory bank*, and therefore it is required to select that memory in which the data needs to be written into (or read from). Next the read-write (R/W) signal must be set to '0' (write mode), which will then activate the input buffer. These buffers drive the data along the data bus line and the selected bit lines to the selected memory cell. When the value of 'data in' is logic '1', then this value will be stored in the cell. Next, all word lines and bit lines are deselected, and all stored data will be maintained in the memory cells.

When the memory is being read, it must first be selected through the CS signal again. Then the R/W signal must be set to read ('1'). After the row and column addresses have been supplied to the address inputs $A_5A_4A_3A_2A_1A_0$ and $A_{11}A_{10}A_9A_8A_7A_6$, respectively, the data stored in the selected cell is then forwarded through the bit lines and the data bus lines to the sense amplifier. Because a memory cell needs to be very small, its signals are very weak, and a very sensitive amplifier is therefore used to quickly detect whether there was a '1' or '0' in the cell. Having set the R/W signal to '1' (read mode) has also activated the output buffer, which will drive the sense amplifier output signal to the output pin (*data out*). This output may be connected to an on-chip bus line, when the memory is embedded on a chip, or to a bus line that communicates with other chips on a PCB, in case of a stand-alone memory chip. A 4 kb, as shown in Fig. 6.5, only included one sense amplifier. Today's memories have a sense amplifier in every column.

The above-described memory write and read cycles are approximately the same for all memories, with the exception that the write cycle in a non-volatile memory is much more complex. The main focus in the following subsections is therefore on the architecture and operation of the basic memory cells, as they greatly determine the properties and application areas of the different memories.

6.3 Volatile Memories

6.3.1 Introduction

There are many applications that need memories for fast storage and retrieval of data. For these applications, RAMs are the perfect solution because they allow write and read operations at very high frequencies. They are usually located close to the (micro-) processor cores that perform fast operations and need to quickly exchange data with these memories. There are two categories of volatile memories: static random-access memories (SRAM) and dynamic random-access memories (DRAM). Their operation, density and applications are quite different, and so we will summarize them in the following discussions.

6.3.2 Static Random-Access Memory (SRAM)

The storage principle of the memory cell in a *static RAM* (*SRAM*) is based on the positive feedback of two inverters (transistors T_1 to T_4 in Fig. 6.6).

During writing into the cell, first the data is put onto the bit lines, as described in the previous subsection. When the word line is selected (high), the cell is connected to the bit lines by T_5 and T_6, respectively, and the state of the bit lines will be copied

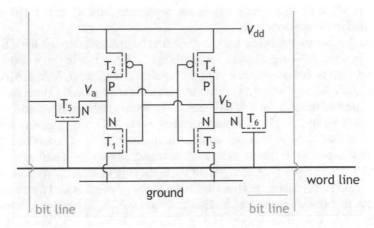

Fig. 6.6 Six-transistor static RAM cell with additional bit line precharge transistors

into the cell. T_1 to T_4 represent the memory cell, which will then maintain the data by the two inverters forming a positive feedback loop. Once the data is written into the cell, the word line will go low, and the data is electrically locked in the cell and will remain there until the cell is rewritten or when the supply voltage is switched off. During reading, the bit lines are first switched to V_{dd} and then kept floating high. Next, when the word line goes to high level again, the state of the cell is copied onto the bit lines.

Each SRAM memory cell requires at least six transistors (6-T cell; T_1 to T_6 in Fig. 6.6). An SRAM is therefore relatively large, and not so many cells do fit on one single chip.

As transistor sizes and voltages have scaled dramatically over the years, 6-T SRAM cell operation is becoming less reliable and alternatives like 7-T or 8-T cells are being considered [4].

Because SRAMs are built from transistors only, their fabrication is compatible with that of digital CMOS circuits and can thus easily be embedded on a system chip. Almost all embedded random-access memories on a logic chip (e.g. for all mobile, automotive, consumer, computer, applications, etc.) are therefore implemented as SRAM.

6.3.3 Dynamic Random-Access Memory (DRAM)

In applications that need much more memory capacity (several GB), the memory is usually implemented as a *dynamic RAM* (*DRAM*). The storage principle is quite different from an SRAM. Each memory cell in a DRAM contains only one transistor (T) and a capacitor (C), which can store a certain amount of charge (Fig. 6.7). It is therefore also called 1-T memory cell.

A logic '1' is written into the cell by completely filling the capacitor with charge (or filling the 'pond' with water, in the analogy in the right figure). A logic '0' is

Fig. 6.7 (a) Circuit diagram of a DRAM. (b) Water model of a DRAM cell

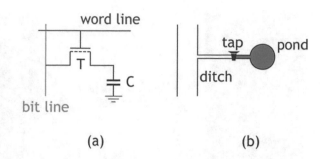

(a)   (b)

represented by an 'empty' capacitor (empty pond). But, these capacitors are not ideal. They suffer from a continuous temperature-dependent leakage by the access transistor T. The higher the temperature, the more charge is leaking away from the cell, which is similar to the evaporation of water in the pond. Therefore, each memory cell needs to be read and rewritten within a certain period of time (called *refresh period*), to refresh its data, analogous to the regular refill of the pond during summer. As a result, the operation of a DRAM is more complex than that of an SRAM, but, due to the smaller cell size, it can store 10–20 times more data on the same silicon area. Because the cells are relatively small, they cannot contain much storage charge. The currents involved in a DRAM read operation are therefore smaller than in an SRAM causing longer access and cycle times. The DRAM vendors do everything possible to increase the memory cell capacitor that stores the charge. But, scaling the cell dimensions normally results in a smaller cell storage capacitance. Therefore, in analogy to a deeper instead of a larger pond to increase the water volume, they have decided to use the third dimension, the height, to increase the capacitor volume, without increasing the cell area.

Figure 6.8 shows a cross section of a *stacked-capacitor cell* and a *trench-capacitor cell*. The stacked or trench capacitors are fabricated far above or below the normal transistor layers, respectively. Because of that, they require several complex fabrication steps and are made in special memory fabs with specific fabrication tools to create the complex 3D stacked-capacitor cells. These process steps are quite different from those used to fabricate normal integrated circuits. DRAMS are therefore mostly realised as stand-alone memory chips. Because trench cells face increasing stress from within the substrate, all DRAM vendors have currently moved towards stacked architectures [5].

Stand-alone DRAMs are therefore fabricated in production processes that are completely tailored to their memory architectures, to achieve the best memory performance and the highest yields.

In summary: compared to DRAMs, SRAMs offer much less memory storage capacity but at faster read and write cycles. The basic idea behind the cache memory (see Fig. 6.1) is that it combines the speed of a small but expensive memory (usually SRAM) with the large storage capacity of a slow but cheap memory (usually DRAM). Today, large parts of the cache memories are often embedded on the processor implemented as SRAMs. The 24-core Intel® Xeon® Processor E7-8890 v4

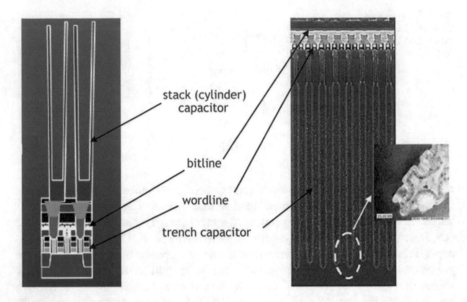

Fig. 6.8 Example of using the third dimension to increase the memory cell capacitor: a stacked-capacitor cell (left) and a trench-capacitor cell (right). (Source: former Qimonda)

has 60 MB (=480 Mb) of embedded cache memory. The capacity of the main memories of computers and laptops (realised with a number of stand-alone DRAM memory chips) currently ranges from 4 to 32 GB. These are stand-alone memories. They are usually expandable, by plugging additional DRAM chips into the spare memory bank on the motherboard of a computer (see Part I, Fig. 3.14). As stated in the introduction, the memory capacity of the cache and main memories are crucial to the overall performance of a computer system, because they determine how many programs can run simultaneously and how much data is available to them, in time.

It is clear, now, that RAMs are used in applications where the data usually needs to change very frequently. However, there are many applications that require very long data storage cycles (days, weeks, months). These applications use so-called non-volatile memories, which will the subject of the next paragraph.

6.4 Non-volatile Memories

6.4.1 Introduction

Since their introduction in the early 1970s, non-volatile memories have become key components in almost every electronic system. As stated in the introduction (Sect. 6.1), *non-volatile memories* are memories that must maintain their data close to 10 years, even when they are disconnected from the supply. Initially, the market was relatively small and mainly driven by mobile applications. During the last decade, however, the number of non-volatile memory applications has increased tremendously, which has made the non-volatile market the second largest after DRAMs.

Popular examples are the storage of pictures in a digital camera, the storage of music in MP3 players, the storage of maps in GPS systems and the storage of digital data on memory carriers such as memory cards and USB memory sticks. These memory carriers are also often used to transfer data from one electronic device to another, e.g. from a digital camera to a PC, or as backup for important data files that are not allowed to get lost. Today, non-volatile memories with USB connection, called solid-state drives (SSD), are increasingly replacing hard disc drives in modern PCs and laptops.

Non-volatile memories include ROM, PROM, EPROM, EEPROM, flash memory, FRAM, MRAM and PCM. In the following paragraphs, the basic architectures of the most popular ones are discussed in some detail, including their basic properties.

6.4.2   Read-Only Memory (ROM)

A *read-only memory* (*ROM*), also known as mask-programmable ROM, is in fact a random-access memory, in which the information or program is written during the manufacturing process. The information is therefore lasting and non-volatile. It can be read but it can never be altered. With the exception of the write facility, the architecture of a ROM is similar to that of a RAM (Fig. 6.5). Different processing layers could be used to store information in a ROM. In many ROMs, today, one of the top metal layers is used to define the '1's and '0's in the memory. Stand-alone ROMs with different memory capacities are already fabricated up to these final metal layers and are therefore available off the shelf. After specification of the storage data, the preprocessed wafers are fed again into the clean room to fabricate these final layers. This will reduce the total processing time to only a few weeks.

Figure 6.9 shows an example of a conventional ROM, which clearly shows a pattern of these '1's and '0's.

ROM memories are used in applications with extremely high volumes, e.g. video games, TV character generators to display on-screen text, etc.

In certain applications, however, the development engineer wants to define his own program to be permanently stored in the memory. These applications require a so-called programmable read-only memory, which is the subject of the next paragraph.

Fig. 6.9 Example of a (conventional) ROM, which clearly shows the patterns of '1's and '0's. (Source: NXP Semiconductors)

6.4.3 Programmable Read-Only Memory (PROM)

A *programmable read-only memory* (*PROM*) is a one-time programmable read-only memory, meaning that it can be programmed only once by the user. Each memory cell contains a fuse, which can electrically be blown during programming. Logic '1's and '0's are represented by blown or not-blown fuses. Modern PROMs use poly-fuse cells in standard CMOS technology. Such memories are also called one-time programmable (*OTP*) memories. The development of PROMs move towards 3-D dimensions with four to eight layers of memory arrays stacked on top of each other. Memory cells are then located between two successive metal layers and positioned at the crossroads of the metal wires in these layers, which run in perpendicular directions. By applying a large electrical field across the cell (by selecting the corresponding metal tracks in the two successive layers, between which the cell is located), its physical state changes, causing a dramatic reduction of its resistance. The cells that are not programmed maintain their high-resistive state. So, the '1's and '0's are represented by the high- and low-resistive states of the respective memory cells. Since these cells are only fabricated between layers above the silicon, each cell could only be $4F^2$ in size, with F being the minimum feature size, e.g. 45 nm in a 45 nm CMOS technology. Still, this memory did not become a commercial success because it could not compete with the low price per bit of NAND flash memories, which will be discussed later in this chapter.

The wish for rewritability in many applications has increased the demand for erasable architectures. These are discussed in the following sections.

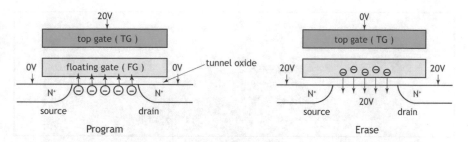

Fig. 6.10 Schematic representation of a floating-gate cell, which is the basis of a non-volatile memory

6.4.4 Erasable Programmable Read-Only Memory (EPROM, EEPROM)

Figure 6.10 shows a schematic representation of the basic reprogrammable non-volatile memory cell and how it is programmed and erased. It consists of one transistor only. However, this is not a normal transistor. Compared to the transistor shown in Fig. 5.1, the transistor used in the memory cell has a 'floating gate' between the (top) gate and the channel.

The data in this cell is represented by the presence or absence of charge on this floating gate. By changing the charge on the floating gate, the threshold voltage of the transistor can be changed. A programmed cell (storing a logic '1') has such a high threshold that it can never conduct, while the threshold voltage in an erased cell (logic '0') is so low that it will always conduct. *Erasable PROMs* (*EPROMs*) and *electrically erasable PROMs* (*EEPROMs*) differ in the way the charge is transferred to and from the floating gate, by different voltages on the source, top gate and drain. Moreover, in an EEPROM, each memory cell also has an additional access transistor, next to its floating-gate memory cell. It enables to erase and program these memories on a bit-by-bit basis, meaning that we can just change the contents of one single cell, leaving the others unchanged. However, this additional access transistor causes an EEPROM to be larger than an EPROM. Further details on their operation are beyond the scope of this book.

Figure 6.11 shows an example process cross section of the basic reprogrammable non-volatile memory cell in a conventional technology, which clearly shows the polysilicon top and floating gate.

Every time that such a memory cell is programmed and erased, the electrons are pulled or pushed through the dielectric isolating material between the channel and the floating gate. This flow of electrons damages this *tunnel oxide*, which imposes a limit on the number of times that a cell can be erased and programmed. This is called *memory endurance*.

Figure 6.12 shows an example of an endurance characteristic. The threshold difference between the programmed and erased states enables a few ten thousand to more than a hundred thousand program/erase cycles for the individual cells. Because

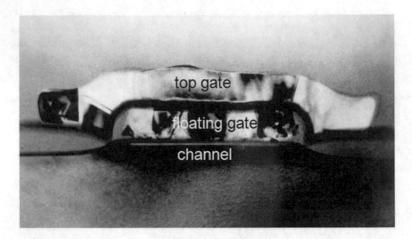

Fig. 6.11 Example cross section of a basic erasable PROM cell

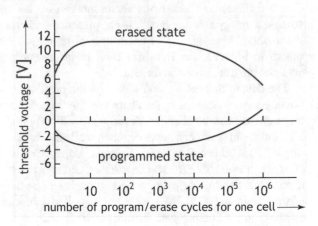

Fig. 6.12 Endurance
characteristic of a
full-featured EEPROM cell

advanced memories contain billions of extremely small memory cells, (E)EPROMs
can be reprogrammed in the order of several thousand to ten thousand times. This
number depends on the different charge storage techniques on the floating gate.

The *data retention time* of almost all erasable PROMS is specified as more than
10 years. This limits the scaling of the tunnel oxide thickness to 7 nm. Below this
thickness, the charge on the floating gate, representing the stored data, would leak
away within the specified data retention time of 10 years.

Erasable PROM applications include conventional consumer applications, universal remote controls, mobile phones and pagers, garage door openers, cameras,
automotive, home audio and video and smart cards. Small versions are also implemented as embedded memories.

Currently, floating-gate devices are also used to fabricate OTPs. Data is then
stored by the absence or presence of charge on the floating gate, which permanently
puts the transistor on or off, respectively. Such an OTP lacks circuitry or other methods to electrically erase the data.

6.4.5   Flash Memories

A *flash memory* is also an erasable PROM in which the complete memory or complete memory sectors (blocks) can be erased simultaneously. Originally, the name flash memory is derived from the original EPROM. This type of memory has a transparent window on top. Erasure is done by exposing the chip to UV light. In fact, during one UV flash, the complete memory is erased.

Flash memories are categorised as NAND flash and NOR flash. All flash memories are arranged into blocks. The typical block capacity is around 128 kB. Erasure cannot be done individually but is done by complete blocks in 'one flash'. The lifetime of a flash chip is determined by the maximum number of program/erase cycles per block. In advanced flash memories, the sizes of and oxide thicknesses within the basic cells are so small that their reliability has reduced significantly compared to the devices of a decade ago. The number of program/erase cycles, based on the physical reliability of the cells, is therefore limited to just a few thousand times. It is therefore crucial that the program/erase cycles are evenly distributed over the blocks. Today's flash memories include control circuitry to perform such a distribution. This means that after erasing all pictures from a flash memory card in a digital camera, the new pictures are not stored on the 'just cleaned' memory positions but on those that were still not used or that were used less often.

The endurance of flash memories is typically specified between 1000 and 10,000 cycles, depending on the application area. Due to scaling and including more bits in one cell, the operating margins reduce. Therefore, many advanced non-volatile memories use an *error-correction code* (*ECC*) and wear levelling algorithms to increase lifetime. ECC is an electronic error detection and correction mechanism, which uses special codes to detect whenever a data bit has gone corrupt and 'repairs' it electronically.

Each individual cell in a *NOR flash memory* (Fig. 6.13 left) contains its own bit line contact, which makes it relatively large but gives it a relatively fast random access. The figure shows that all NOR cells are connected in parallel resembling a kind of NOR architecture. A cell in a *NAND flash memory* is part of a serial chain (Fig. 6.13 right), in which the cells lack an individual connection to the bit line. Instead, it is connected in between two other cells. It is therefore small, with a cell area of only $4F^2$/bit, which is about half the area of a NOR cell. F is the minimum feature size of the technology node, e.g. $F = 45$ nm for 45 nm CMOS technology. This compares with $6F^2$/bit for a DRAM cell and 80-100F^2/bit for a 6-transistor SRAM cell. A NAND flash has a slow random access (typically 20 μs), but it allows fast sequential access (typically 20 ns) for cells in the same serial chain.

Because of their random-access capability, NOR flashes allow high read throughputs and low latency. They have traditionally been used to store relatively small amounts of executable code in applications such as computers, mobile phones and organisers. Currently they are also used as embedded memory in high-data rate applications. Because of their serial architecture, NAND flashes have longer read access. However, the need for low-cost high-density memory drove the NAND flash

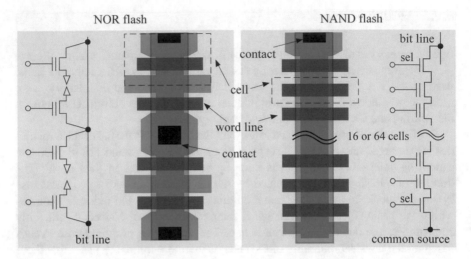

Fig. 6.13 Basic difference between NOR flash and NAND flash architecture. (Source: Samsung)

into the newer mobile devices with increasing performance like mobile phones, MP3 players, GPS, cameras, SSDs and streaming media.

Today's smartphones combine all of these applications in one device, which are commonly supported by NAND flash. NAND flash is also replacing NOR flash in code-storage applications as well. The NAND flash architecture looks similar to the hard disc drive, in that it is also sector based and may also have bad blocks. It therefore requires *error-correction code* (*ECC*) to guarantee correct data retention. The huge capacity of NAND flash modules is a result of the combined advances in memory capacity increase of the individual dies and the increase of the number of dies that can be packaged into a *multi-chip module* (*MCM*). Most *universal serial bus* memory sticks (*USB memory*) and the *memory cards* in digital cameras are using NAND flash memories. They also enable the large memory capacity of *solid-state drives* (*SSD*). The first SSD began to appear in mini laptops and notebooks in 2007, where it replaces the hard disc drive. Advanced SSDs (2018) have a capacity of more than 1 TB or more [6]. All tablet computers have an SSD drive. Compared to hard disc drives (HDDs), SSDs are roughly four to six times more expensive.

Today, particularly the high-capacity 3-D flash memories are also used for cloud storage, where the number of program-erase cycles is very limited.

In all erasable PROM and flash memories the minimum thickness (most commonly 7–8 nm) of the dielectric layers above and below the floating gate is determined by its accumulated dielectric leakage of charge over the specified data retention time (usually >10 years). This has limited the scaling of the vertical dimensions in these memories also resulting in a limited scaling of the program/erase voltages. This leaves some additional operating margin, which can be used to store more bits in one cell to further reduce the cost per bit. In such a *multilevel cell* (*MLC*), different amounts of electron charge can be transferred to the floating gate during programming. Today, many of these chips store two or three bits per cell, which requires four or eight different charge levels on the floating gate, respectively. During a read cycle, the current through the cell is inversely proportional to the amount of charge on the floating gate. This current is measured by very sensitive amplifiers, which each compares the cell current with the current from reference cells. The output level of these sense amplifiers is a measure for the stored bits. Multilevel storage has been known for quite some time. The first commercial multilevel flash memory products were announced at the end of 1996. Another example of a multilevel NAND flash memory is the 4-bit (16-level) 64 Gb [7], fabricated in 45 nm CMOS, which stores four bits per memory cell!! In a *multilevel memory*, the distance between adjacent threshold voltage charge distributions on the floating gate is becoming very small and may lead to a decrease in reliability with respect to read operation and data retention. Therefore, a multilevel flash memory may allow only one to a few thousand program/erase cycles per physical sector, while a conventional flash memory was capable of a ten to hundred thousand of these cycles. Also here, the use of on-chip error-correction coding (ECC) alleviates these problems. The flash memory has penetrated many markets, which were previously dominated by magnetic and optical discs, ROMs, EPROMs and EEPROMs and hard disc drives. The ability to continuously increase the density of flash memories will further continue that penetration.

Because of the continuous scaling of the planar floating-gate memory devices, the space between individual cells became so small that during the programming of a cell, also the charge on the floating gates of its neighbours got affected. This caused a change in their threshold voltages and led to corrupted bits. So, next to using more electrical levels in the memory cell to increase the bit density of flash memories, also multiple layers of stacked memory cells were introduced. Figs. 6.14

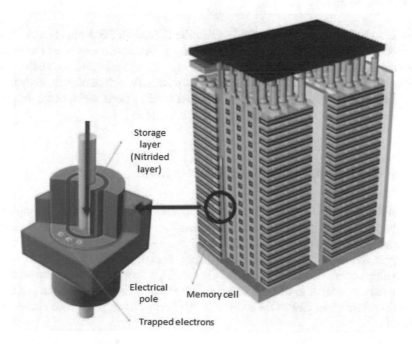

Fig. 6.14 Drawn cross section of 3-D stacked strings of NAND-flash cells. (Source: Toshiba)

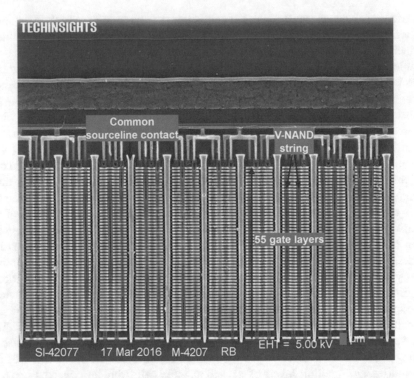

Fig. 6.15 SEM photograph of a vertical cross section of the Samsung 48L V-NAND Array. (Source: TechInsights)

and 6.15 show a drawn cross section (Toshiba [8]) and a SEM photograph of 3-D stacked strings of NAND flash cells (Samsung [9]), respectively. These additional memory array layers only require a limited amount of mask and processing steps and only support the fabrication of the memory cells. Selection of these cells is also performed by the selection circuits located in the bulk silicon wafer. More details on 3-D non-volatile memories technology can be found in [2].

6.4.6   Non-volatile RAM and Battery RAM

A *non-volatile RAM* (*NVRAM*) combines SRAM and EEPROM technologies on one single chip. This kind of memory is sometimes called a *shadow RAM*. Read and write actions can be performed at the speed of an SRAM during normal operation. However, the RAM contents are automatically copied into the EEPROM part when an on-chip circuit detects a power supply dip. This operation is reversed when the power supply returns to its normal level. An *NVRAM* therefore combines the data retention time of an EEPROM with the high-performance read/write cycles of an SRAM.

A *BRAM* comprises an SRAM and a battery, which provides sufficient power to retain the data when the memory is not accessed, i.e. when the memory is in the standby mode. The battery is only used when the power is absent. An SRAM is chosen because of its low-standby power consumption. The battery is included in the BRAM package, and the data retention time is close to 10 years. This type of memory is also called *battery backup RAM*. These 'non-volatile RAMs' are relatively expensive, but then they are used in applications that require data retention even after the power is lost. Such applications include medical, aviation and military. This technology is still available but only works for a limited time. Another disadvantage is that the batteries occupy most of the available space.

6.4.7 *Alternative Non-volatile Memories and Emerging Technologies*

One of the problems related to the scaling of floating-gate devices is the relatively large minimum thickness ($\approx$7 nm) of the dielectric isolation layers above and below the floating gate. This is required to limit charge leakage from the floating gate through the surrounding dielectrics to guarantee a sufficiently long data retention time. There are several alternative memory solutions that are currently in low-volume production. Among these are the magnetoresistive RAM (MRAM) and ferroelectric RAM (FRAM), but so far, these memories have not shown a cost-competitive edge over the existing non-volatile memories.

A lot of research effort, however, is devoted to develop the Holy Grail: a *universal memory* that has the fast write and read capability of random-access memories combined with the non-volatile capability of the erasable PROM and flash memories.

Several alternatives are currently in development and/or in small-volume production. An interesting non-volatile RAM alternative is the so-called phase-change memory (*PCM*), also known as *PRAM*, *PCRAM* and *Ovonic Unified Memory* (*OUM*). Its basic operation uses a unique property of polycrystalline chalcogenide alloy. This so-called phase-change property is also used for recording and erasing in optical media (rewritable CD, DVD and blue ray discs). In these media the required heat for programming is generated through exposure to an intense laser beam. Figure 6.16 shows a cross section of a basic PRAM storage cell.

Under the influence of heat, the polycrystalline state can be changed into an amorphous state and back. Only a small programmable volume of the material is locally heated to above its melting point. This heat can be supplied by a current pulse through the heater. When rapidly cooled, the chalcogenide material is locked into its amorphous high-impedance state. By heating the material above its crystallisation but below its melting temperature, the cell switches back to its low-impedance crystalline state. The difference in impedance between the two states is between one and two orders of magnitude. During a read operation, a voltage is

Fig. 6.16 Basic cross
section of a phase-change
memory cell

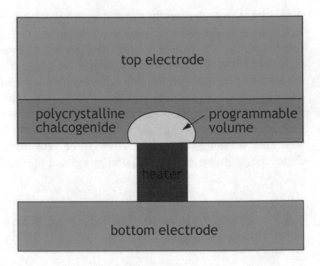

applied to the cell, and the read current is measured against the read current of an
identical reference cell with a fixed logic state. Today's PCRAM complexity is still
far behind that of DRAM and SRAM, but due to the extending application areas
with limited power budgets, particularly in hand-held devices, there is more pressure to further develop this real non-volatile RAM. Volume production of the first
commercial PCRAM was announced for the second half of 2007. Examples of
PCRAM designs can be found in [10]. PCRAMs are currently seen as the most
promising successor of NOR flash, particularly in hand-held phones, where it is
expected to extend battery life with more than 20%.

An alternative non-volatile memory technology is the 3D XPoint™ memory
[11]. It is being co-developed by Intel and Micron Technology and built from a
transistorless architecture in which the cells are created between two successive
metal layers at the intersection of word lines and bit lines. As such, each cell can be
individually written (programmed and erased) and read at much faster rates than ash
memories, as these only enable block-level erasure and programming. 3D XPoint™
is not seen as a replacement technology for either NAND ash or DRAM. The endurance is expected to be a few million program/erase cycles, which is roughly a thousand times higher than that for NAND ash. Figure 6.17 shows a cross section of this
cross-point memory. The cell consists of two parts: an ovonic switch to select the
cell in series with a material-property change storage element, which changes the
particular property over the full bulk of its material. It may use chalcogenide materials (although not fully confirmed by the inventors) for both selector and storage
parts. This is claimed to increase scalability, stability, endurance, robustness and
performance. It can have a low-resistance or a high-resistance state, depending
whether there was a logic 'one' or logic 'zero' stored (programmed) in the cell. Full
details of the technology have not yet (2016) been given, but it is claimed to be no
phase-change nor memristor technology [12].

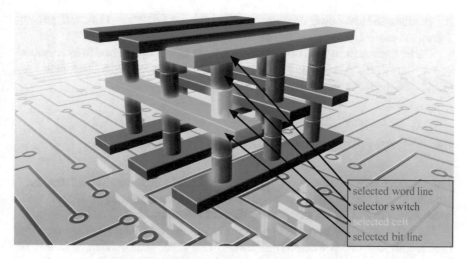

selected word line
selector switch
selected cell
selected bit line

Fig. 6.17   Cross section of Intel/Micron 3D XPoint memory. (Source: Micron Technology)

A first-generation 128 Gb is currently being fabricated by Intel and Micron in a 20 nm two-layer process in a jointly operated fab on sample basis. This new technology is not expected to replace existing technologies but will find its applications in between those of NAND and DRAM.

Many other non-volatile memory techniques are currently in basic R&D phase and still need a lot of engineering and development time, before they really become a mature high-volume product, if they ever will.

6.5   Memory Interfaces

Today's multi-core (dual-core, quad-core, eight-core, twelve-core, 24-core, etc.) processors run at frequencies of several gigahertz. Figure 1.11 (Part I) shows the AMD eight-core Ryzen 7 1800X die, on which eight identical cores together can perform the operations many times faster or execute a lot more tasks in parallel. To be able to fully explore the benefits of this processing power, also the communication between the processor and the memory chips must be fast. Although large parts of the memory are embedded on the processor chip, the main memory is still a separate device, which requires very fast read and write accesses.

Communication with large external memory is mostly performed through special memory interfaces. A very popular memory interface is the *DDR interface*. DDR is the abbreviation of *double-data rate*, which enables doubling of the data rate without increasing the clock frequency. This is done by transferring two data samples, instead of the usual single data sample, per clock period. There are several generations of DDR (e.g. DDR 2, DDR3, DDR4, etc.), each with its own voltage level and frequency of communication. There are even special memory interfaces for graphics and low-power applications, respectively, GDDR and LPDDR. Chapter 7 presents more details on several of these types of interfaces.

The frequency (or speed) of communication is generally referred to as *communication bandwidth*. Also in the future, the memory interface bandwidth remains one of the limiting factors in the overall performance of a system.

6.6   Stand-Alone Versus Embedded Memories

The integration of a complete *system on a chip* (*SoC*) includes the combination of analog, digital and memory cores. Usually this is done to achieve higher speed, smaller area or less power or a combination of these three.

Basically, there are three different approaches to implement a system on a chip (SoC). The first one is to embed memories in a logic chip, fabricated in a logic-based fabrication process (*embedded memory*). Basically all common types of memories can be embedded: SRAM, DRAM, ROM, E(E)PROM and flash memories. Except for SRAM, they all require several additional masks and processing steps to efficiently embed them on a SoC. Traditionally, the advantage of *stand-alone SRAMs* was twofold: they offered a higher speed than DRAMs and consumed less power in standby mode. Both advantages, however, are losing ground. The speed of the SRAMs has become so high that their usage is more and more hampered by the limited bandwidth of the interface between ICs. Reducing operating margins in combination with increasing SRAM leakage may turn a read operation into a parasitic write, which may flip the cell during reading. This leads to almost conflicting transistor sizing requirements for the read and the write operation. The increasing leakage is not only a limiting factor in achieving low-standby power levels, it also contributes to reduced operating margins of the basic SRAM cell. This has resulted in a decrease of interest and production of high-density stand-alone SRAMs. This is also reflected by the papers in the SRAM sessions at the major chip conference (*ISSCC conference* [13]) over the last decade. The focus has shifted from high-capacity stand-alone Gb SRAMs to high-performance high-density on-chip caches in high-performance CPUs and alternative SRAM cells with improved robustness.

Compared to stand-alone memories, embedded memories show a much lower bit density. The difference between the bit density of an embedded RAM (SRAM) and the bit density of a stand-alone RAM (DRAM), in the same technology node (not in the same technology!!), may be in the order of a factor ten to twenty. Embedded SRAM can be found in practically every application. Also the usage of embedded flash spans a wide range of applications, such as microcontrollers, industrial, office automation, networking, automotive, consumer, mobile applications, smart cards and RFID tags. Today, the increasing system requirements, expressed in the growing number of features and bandwidth, have led to a growth of both the capacity and number of distributed embedded memory instances on a single chip. The high-performance microprocessors of Intel and AMD incorporate level 1, level 1 and even level 3 caches and contain several billion transistors, of which most (often more than 70%) are located in these embedded memories.

Testing is a problem that arises with the merging of huge memory blocks with logic on a single chip. In Chap. 12 we will come back to the testing of memories as well.

6.7 Classification of the Various Memories

Table 6.1 includes a summary of the most important parameters that characterize the different memories. It shows the difference in memory sizes but also some performance parameters, such as memory access and cycle times. Parameter F in this table refers to the minimum feature size in the corresponding technology node, e.g. 45 nm in the 45 nm node, etc.

The numbers in this table are average values and representative for stand-alone memories and may vary between different memory vendors. The characteristic values of these parameters render each type of memory suitable for application areas, which are summarised in Table 6.2.

Table 6.1 Characteristics of different stand-alone memories

Devices		SRAM	DRAM	ROM	PROM	NOR-flash	NAND-flash	FRAM
Physical cell size		$150-200F^2$	$4-8F^2$	$4F^2$	$4F^2$	$8-10F^2$	$4-5F^2$	$15F^2$
Capacity [bit]		<1 G	<32 G	<1 G	<64 G	<16 G	<512 G	<1 G
Max. number of programming cycles		∞	∞	1	1	10^4-10^5	10^3-10^4	$10^{10}-10^{12}$
Programming time (write)		5–40 ns	20–100 ns	–	5–80 ms	5–10 µs	100–300 µs	80–120 ns
Access time (read)		5–20 ns	10–70 ns	5–20 ns	5–20 ms	random: 80–150 ns serial: 80–120 ns	random: 10–20 µs serial: 5–50 ns	80–120 ns
Retention time	no power supply	0	0	∞	∞	>10 year	>10 year	>10 year
	power supply	∞	2–64 ms					

Table 6.2 Application areas for the various stand-alone memories

SRAM	Super-fast systems, low-power systems, cache memories in PCs, workstations, telecommunication, multimedia computers, networking applications, mobile phones, supercomputers, mainframes, servers, embedded memories
DRAM	Medium to high speed, large computer systems, desktop, server, low-cost systems, networking, large volumes, PC, hard disk drives, graphics boards, printer applications, PDAs, camcorders, embedded logic
FRAM	Low-power, non-volatile applications, smart cards, RF Identification, replacement of non-volatile RAM and potentially high-density SRAM
ROM	large volumes, video games, character generators, laser printer fonts, dictionary data in word processors, sound source data in electronic musical instruments embedded memories
EPROM	CD-ROM drives, modems, code storage, embedded memories
EEPROM	Military applications, flight controllers, consumer applications, portable consumer pagers, modems, cellular and cordless telephones, disk drives, printers, air bags, anti-lock braking systems, car radios, smart card, set-top boxes, embedded memories
FLASH	Portable systems, communication systems, code storage, digital TV, set-top boxes, memory PC cards, BIOS storage, digital cameras, PDAs, ATA controllers, flash cards, battery powered applications, mobile phones, embedded memories, MP3 players, e-books, tablets
NVRAM BRAM	Systems where power dips are not allowed, medical systems, space crafts, etc, which require fast read and write access

6.8 Memory Yield

Embedded memories are the highest-density circuits on a chip. Stand-alone memories show even higher transistor densities. This is the reason that these circuits often have the lowest yield per unit of area. Particularly at the introduction of a new stand-alone memory, the yield may be as low as only a few percent. Therefore most memories, including many embedded memories, have on-chip redundant rows or columns. This *redundancy* is further discussed at the end of Sect. 12.3.2.

6.9 Conclusion

The MOS memory market turnover currently represents about one third of the total chip market. This indicates the importance of their use in various applications. These applications have different requirements with respect to memory capacity, reprogrammability, retention time, power consumption, access time, bandwidth and data. Modern integrated circuit technology facilitates the manufacture of a wide range of memories that are each optimised for one or more application domains. The continuous drive for larger memory performance and capacity leads to ever-increasing bit densities although limits are appearing on the horizon. The DRAM and flash markets show the largest volumes and, not surprisingly, the highest demand for new technologies. This has resulted in the presentation of the first 64 Gb flash memory at the 2009 ISSCC conference. The first flash memories have already shown up as *solid-state drives* (*SSD*) in laptops and tablet computers, and it is only a matter of time before the cost per bit of non-volatile semiconductor memories makes them an attractive alternative to magnetic and mechanic hard discs in other high-volume computer applications as well.

This chapter gave an overview of the basic operating principles of the most popular range of embedded and stand-alone memory types.

The characteristic parameters of the stand-alone types are compared in Table 6.1, and their application areas are summarised in Table 6.2. The memory capacity values (number of storage bits) in the table refer to high-volume memory products in the year 2018.

Chapter 7
Analog IP, Interfaces and Integration

7.1 Introduction

The title of this book implies a focus on the digital contents of an integrated circuit. This does not mean that analog circuits are no longer of interest. On the contrary, the performance of the analog circuits and chip interfaces may be the differentiating factor to achieve the best overall system performance. A short summary on these circuits is therefore a minimum requirement to complete the picture of integrated circuits.

As stated in Part I, Sect. 3.3, all signals that we feel, see or hear are analog by nature. You can feel pressure, you can hear sound and you can see brightness. The intensity of this pressure, sound and brightness can vary over a wide range of continuous values. Although we currently see a lot of advanced electronic applications like digital TV, tablet computers, smartphones, e-readers, GPS, car info systems, smart cards, etc., they always need to interface with analog sensors, devices or circuits, because the world is analog. In most of these applications, an IC executes only part of the functionality of the total system. It performs analog, digital, storage and mixed-signal operations on a variety of input signals (chip inputs) and creates analog and/or digital results, in the form of output signals (chip outputs). To exchange data with other parts of the system, it uses so-called interfaces and *input and output (I/O) terminals*. A lot has already been said about the difference between analog and digital circuits. Because analog circuits closely interact with the real world through a variety of interfaces, they are described in this chapter in combination with examples of these interfaces and I/O pads.

Together with Chaps. 5 and 6, this chapter completes the description of the various chip components (digital, memory, analog and interface circuits). Therefore, a summary of the integration of these components onto a single chip is included in the final paragraph.

© Springer International Publishing AG, part of Springer Nature 2019
H. Veendrick, *Bits on Chips*, https://doi.org/10.1007/978-3-319-76096-4_7

7.2 Analog Circuits and IP

The simplest CMOS circuit representation is that of an inverter (Fig. 7.1).

When used as a digital circuit (see Chap. 5), its input and output signal levels can only have two discrete values: low (ground) or high (supply voltage V_{dd}). When the input is at ground level (0 V), the output will be at V_{dd} level and vice versa. An analog circuit can also operate on all signal values in between ground and V_{dd}. When the above circuit is symmetrical in operation, the output will be $\frac{1}{2}V_{dd}$ when its input is at $\frac{1}{2}V_{dd}$. As explained in Chap. 1, a transistor will carry a larger current when it has a larger width. Assume that we now design the nMOS transistor two times wider compared to the previous symmetrical situation, then the output will be a lot smaller than $\frac{1}{2}V_{dd}$ when the input is still at $\frac{1}{2}V_{dd}$. In other words, the same small-signal input swing will result in a larger output swing, and as such, the circuit has a larger amplification. So, in a digital circuit, we are only interested in the end result of the output signal: the output is either low or high, while in an analog circuit we are interested in the complete shape of the signal.

Therefore, the design of *analog circuits* has always been and will always remain high-precision work at the device and circuit level by specialists. The following IP cores belong to the most commonly used analog blocks on a chip. Only a glimpse of their functionality is presented here. A description of their circuit implementation can be found in many books on analog integrated circuits but is beyond the scope of this book.

Analog-to-digital (A/D) converters and digital-to-analog (D/A) converters are almost used in every system. As its name implies, an *A/D converter* converts analog signals to a digital domain (binary format) to enable digital processing and storage of the data. A *D/A converter* is used to convert the digital data back into analog signals, after the digital operations have been completed. Depending on the number of bits, the clock speed and process technology node, the chip area of these converters may vary from less than one to a few square millimetres. A summary on A/D conversion is given in Sect. 3.5 in Part I.

Analog filters are circuits that pass or block certain frequency ranges of analog signals. As such, the circuit has a frequency-dependent transfer function. A simple *low-pass filter* only transfers the low-frequency components of a signal, while atten-

Fig. 7.1 Basic circuit
schematic of an inverter

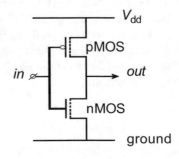

uating (blocking) its higher frequencies. A *band-pass filter* only passes a certain range of frequencies (*bandwidth*), while it blocks frequencies outside that range. An example of such a filter is the tuner in a radio receiver, which passes the frequency of the required broadcasting station, while blocking the frequencies of the unwanted stations. Analog filters are also used as data conversion filters to reconstruct the correct analog signal. Conversion of digital circuits back to the analog domain often introduces unwanted frequency components generated by the sample frequency and its harmonics. These need to be filtered away. Filter die areas depend on application and technology node and may vary between 1 and 10 mm^2.

Amplifiers are circuits that increase the power (strength) of a signal. A well-known example is the audio amplifier which amplifies the extremely small input signal (e.g. from a microphone, receiver or DVD player) to such a level that it can stimulate a loudspeaker. A general problem of an amplifier is that it does not equally amplify all frequencies in the input signal. Another problem is the noise that is present in the original small signal but also the noise that is generated during the amplification process. An additional concern is the dynamic range of the amplifier, which represents the range between the smallest and largest output levels, in which noise and distortion levels are still acceptably low. These and other parasitic effects increase the design challenges of a high-quality amplifier.

RF circuits are used to build the necessary transmitter and receiver circuits needed in the rapidly increasing radio and wireless applications. The basics of wireless communication have already been discussed in Sect. 3.6 in Part I.

Several analog circuits are used to synchronise signals with a reference. An example is a *phase-lock loop* (*PLL*), which extracts the clock signal from the data signals in a receiver to track one system with another one. A well-known application is FM radio, where it locks its frequency to the carrier signal frequency (see Fig. 3.11 in Part I). PLLs are also used as *frequency dividers* or *frequency multipliers* to generate clock signals with frequencies that are fractions or multiples of the frequency of the input signal, respectively. These clock signals are used to synchronise the communication in complex systems that may consist of multiple clock domains.

Finally, *analog interfaces* realise a wide range of communication applications from chip-to-chip interfaces (I/Os) to board-to-board interfaces. The large variety of electronic systems and products require different interface solutions, each with a dedicated optimal protocol, bandwidth and speed. To meet the increasing bandwidth requirements, many applications experience a transition from parallel to high-speed serial I/Os. Below, in Sect. 7.3, a short summary on interfaces provides only a flavour of the fast-growing interface solutions.

During the 1970s, before the development of digital cores and advanced computer-aided design (CAD) tools really took off, most ICs contained only a simple basic analog circuit, like an operational amplifier, which was built from only a few transistors. The designers at that time applied relatively simple calculations and manual layout design styles. Because analog circuits are very sensitive to process and environmental variations, a lot of 'overdesign' was used to guarantee correct operation. Proven designs were incorporated in a kind of library, which then enabled a gradual increase of IC complexity by reusing these library blocks. With the advances in computer performance, software programs were developed, which allowed the simulation of more complex circuits with much greater accuracy than was possible with 'hand design'. One of the first analog circuit simulators was SPICE, which enabled the designer to accurately simulate the function and performance of analog and mixed-signal circuits. Because such a simulation includes both a complete description of the operation of all devices including all of their parasitics, it could only be used for circuits with a limited complexity.

Analog and mixed-signal circuit simulation tools have been continuously improved since then, and today several tools are included in a so-called analog/mixed-signal (AMS) design flow. While a digital synthesis tool can automatically transfer a wide range of high-level specifications into a digital circuit implementation, the use of analog synthesis is limited to only well-defined parameterisable architectures like operational amplifiers and filters. Moreover, analog circuits are closely communicating to the dynamic world of sound, light, pressure, temperature, voltage, etc. This may require a broad scope of specifications with tight margins. Therefore, the design of *analog IP* (intellectual property) remains much more cumbersome than digital IP design. Because many advanced CMOS technologies offer additional analog and RF process options, which enable the integration of a complete *system on a chip* (*SoC*), most IC vendors, today, offer more and more analog IP as well. To speed up the design of mixed-signal circuits, the IC vendor must have sufficient analog IP available in their libraries in order to allow relatively quick and easy integration of the analog parts for both high-speed and low-power mixed-signal applications.

As mentioned before, analog circuits are very susceptible to process and environmental variations. Therefore, analog IP must be verified for different products and different foundries. Also the portability to other technology nodes is an important issue. A summary on analog circuit and IC design characteristics can be found in [14].

The continuous scaling of the physical transistor sizes, together with the reduction of the supply voltages close to or below 1 V, had a dramatic impact on the operating margins and performance of analog circuits.

A mixed-signal chip (Fig. 7.2), today, may contain various IP: memory (RAM, ROM, flash), digital (application and glue logic, DSP, microprocessor), analog, mixed-signal, RF and interfaces.

Such a chip usually has a couple of different supply voltages. The digital parts then run at the nominal supply voltage, while most analog parts may run at higher voltage levels in order to keep sufficient operating margins. Many analog circuits also use larger than minimum transistor sizes to reduce the sensitivity to process parameter spread. Also the interfaces and I/Os may operate at different voltages, dictated by the target application.

Chapter 3 in Part I includes introductions on A/D and D/A conversion and RF systems. A lot more could be said about the design of analog and mixed-signal IP but is beyond the scope of this book.

Fig. 7.2 Example of a
mixed-signal chip

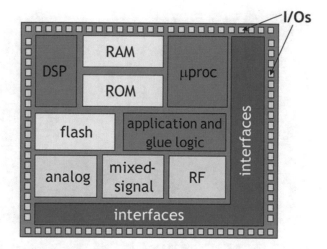

7.3   Chip Interfaces

Complex mixed-signal ICs may contain a large number of digital IP cores and memory IP cores and a variety of analog IP cores. Particularly the latter group of cores often communicate closely with various interfaces on the chip. These interfaces are usually physically implemented as a number of input/output terminals also called *I/O pads*. Next to these signal I/Os, the chip needs to be connected to one or more supply voltage(s) through a variety of power and ground terminals, also called *power and ground pads*. Their number depends on the maximum amount of current (power) that the chip consumes during full operation. On the silicon die, these I/Os and power and ground pads must be connected to the leads in the lead frame of the package in which they will be mounted. These connections are often realised through very thin wires (*bond wires*). Therefore, most on-chip I/O and power and ground terminals are located in the periphery of the die. An I/O pad serves three basic functions: connection, protection and interface. These three will be discussed in some detail in the following subsections. Particularly the interface is currently a hot topic in system definition and system design. It is therefore that this topic will get most of the attention in this chapter.

7.3.1   Connection

An I/O pad includes interface circuitry between the chip circuitry and the application and a relatively large metal square (*bond pad*) implemented in the top metal layer of the chip. This bond pad is to be connected (soldered) to a bond wire, which is also connected to the corresponding lead and pin in the package. Figure 7.3 (left)

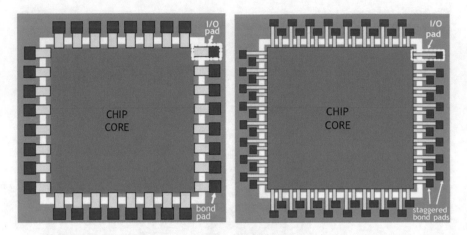

Fig. 7.3 Chip with a single I/O pad ring (left) and a chip with a double row of staggered pads (right)

shows a symbolic representation of a chip with all its functionality inside the chip core and all supply pads and I/Os in the peripheral *pad ring*.

Complex ICs may require so many pads that they don't fit in a single pad ring along the edges of the core of the chip. In this case the pad ring would be much larger than required by the core area, which is known as a *pad-limited design*. Therefore a *double pad ring* with *staggered bond pads* (Fig. 7.3 (right)) has become available. The pad pitch in a double ring is much less than in a single ring, but this is at the price of an increased pad height. Also triple pad rings exist. All peripheral pad rings have their bond pads connected through wires to the package leads (see Fig. 14.3).

When even a triple pad ring does not offer sufficient pads, then an array of pads distributed over the total chip area is used. In this case no wires are used for the connection, but a droplet of solder is deposited on each pad. Then the chip is flipped over and directly attached to an array of balls located on the chip carrier or circuit board and which is an exact copy of the pad array. This *flip-chip* technology and other package-related topics are discussed in Chap. 14.

7.3.2   Protection

The continuous growth in *Internet of things (IoT)* applications led to an almost unlimited number of electronic systems that need to operate reliably in an increasingly complex electromagnetic environment.

I/Os also play a dominant role in the robustness and reliability of an IC. Because these are the external connections to a chip, they need to be physically contacted, which makes them very prone to external *EMI (electromagnetic interference)*. Bond pads, package leads and board wiring act as antennas. They can 'send' or 'receive' an *electromagnetic pulse* (*EMP*), which can dramatically affect the chip's own operation or the operation of neighbouring electronic circuits and systems [15].

Integrated circuits are exposed to many possible sources of damage, both during and after the manufacturing process. The principle cause of damage is *electrostatic discharge* (*ESD*), due to the transfer of charge between bodies at different electrical potentials. ESD pulse durations are very short and normally range from 1 to 200 ns, but they may introduce very large power spikes into the chip. The high impedance of MOS input circuits makes them particularly vulnerable to physical damage when they are exposed to such spikes. This may result from operations during the fabrication process or from handling (un-)packaged dies and during bonding. It may also occur during testing and maintenance or in the application. While only a few devices or connections may be severely damaged, many more may suffer damage that is not immediately apparent. These latent failures will result in *customer returns*, which is one of the biggest worries of semiconductor vendors.

ESD is therefore one of the most important factors that determine the reliability of an IC. The damage caused by ESD is irreversible. The human body is one of the main sources responsible for ESD. Just by walking on a carpet on a low-humidity day, for instance, a person wearing shoes with highly insulating soles can build up a voltage in excess of 30,000 V. The resulting charge can then be transferred via an ESD to an electronic circuit during touching. It is also very important that precautions need to be taken to prevent ESD damage during IC fabrication. In addition, protective measures must be included in the design of an IC to ensure that it can withstand acceptably large EMI pulses. On-chip MOS *ESD protection circuits* are used to increase the immunity of an IC to these EMI pulses. Therefore, each I/O, supply and ground pad contains protection circuitry, which prevents excessive voltages to arrive at the IC's input, output and core transistors. The basic idea behind these ESD protection circuits is simple. They contain diodes that clamp negative voltage spikes to a level equal to 'ground-0.6 V' and positive voltage spikes to 'supply + 0.6V'. These diodes, which are relatively large to drain the large current surges during an EMP, are part of the circuitry that surrounds the bond pad.

In many applications, an IC has to communicate with ICs or other electronic circuits or devices that operate at higher voltages. This means that the chip I/Os usually apply higher voltage levels than the circuits in the core of the chip. During the CMOS fabrication process, also parasitic devices are created. As long as these devices are not active, they do not influence the performance and operation of the chip. However, one of these parasitic devices, called the *thyristor*, may become active (triggered), when the voltage on one of its inputs is high enough to trigger its latch function. When this happens, a positive feedback circuit in the thyristor may indirectly short the power and ground lines locally on the chip, causing huge currents and incorrect chip operation. This latch function is then maintained as long as the voltage is supplied to the chip. This effect is called *latch-up* and is mostly initiated at the chip I/Os, because these often run at a higher voltage, which can trigger the thyristor more easily. Therefore I/O cell designers will pay extra attention to the design of these cells in order to prevent the chip from any latch-up behaviour.

7.3.3 Interface

The I/O, power and ground pins play a dominant role in the performance, area, power consumption and robustness of a chip. First of all, an output has to drive a relatively large load capacitance through the chip and board connections and wires. Second, it needs to do that at very high frequencies, because most of the modern electronic systems and gadgets require very large communication bandwidths. Third, the large number of bits (e.g. 32-bit or 64-bit words) that represent one digital input or output not only affects the total chip area but, in combination with the large capacitance and high frequency, also the power consumption of the chip.

From a hardware perspective, an integrated circuit typically has several ports or buses, which interface to a variety of peripherals such as displays, cameras, memories or communication devices. The selection of a chip-to-chip interface depends on many factors, which include size, supply voltage, power consumption, number of required pads and *latency* (number of clock periods that the signal requires to propagate through the interface). Current IC design library packages include a large variety of standardised I/O interfaces, tailored to the specific application area. A selection of this variety is shown in Table 7.1.

A complete list would be much larger, but the table should give the reader an impression on the large variety of existing interfaces. Most of the interfaces included in the table even have different versions. To present a flavour of these, we will give more details on two frequently used interfaces: DDR and USB.

Table 7.1 Various chip interfaces and their application area

Interface	Meaning	Example application
LVTTL	Low-voltage transistor-transistor logic	Conventional TTL circuits
DDR	Double data rate	DRAM memories in servers
QDR	Qdata rate	SRAM memories
SDRAM	Synchronous DRAM	DRAM memories
SSTL-2	Stub series terminated logic	DDR memory interface
PCI	Peripheral component interface	Internal computer devices
PCI-X	High-performance PCI	Servers
USB	Universal serial bus	Peripheral computer devices
CML	Current-mode logic	Telecommunication
HSTL	High-speed transceiver logic	Telecommunication
LVDS	Low-voltage differential signalling	Telecommunication
XAUI	10-Gb attachment unit interface	Local area network (LAN)
PCI-Express	Serialiser-deserialiser (SerDes)	Graphics cards
SPI-4.2	System packet interface level 4	Telecommunication: voice and data
PMU	Power management unit interface	

Table 7.2 DDR family specifications

Type	V_{dd} [V]	Clock freq [MHz]	Data rate [Mb/s/pin]
SDRAM	3.3	100	100
DDR-1	2.5	100–200	200–400
DDR-2	1.8	200–400	400–800
DDR-3	1.5	400–800	800–1600
DDR-4	1.2	800–1600	1600–3200

7.3.3.1 DDR Interfaces

Communication between an IC and a high-speed memory may require 32-bit wide data interfaces (32 parallel bus lines). This number of bits, combined with the frequency at which these bits can change their state, is called the *data rate* or *bandwidth* of an interface. With the rapid performance increase of signal and microprocessors during the last two decades, the communication speed of the memories was, and still is, lacking behind. To alleviate this problem, so-called *double-data rate (DDR)* interfaces have been developed. Conventional interfaces only exchange one bit per bus line for every tick of the clock. DDR interfaces exchange two bits per tick of the clock. In fact, it now changes its data at both the rising and falling edge of the clock signal, which doubles the data transfer rate. The continuous need for increasing performance has forced the development of several generations of DDR (Table 7.2). Higher clock frequencies and data rates go hand in hand with decreasing supply voltages (V_{dd}). This voltage reduction is necessary to limit the power consumption, which would otherwise double each generation as a result of the doubling of the clock frequency.

DDR interfaces have become very popular and are currently dominating the memory interfaces in many communication, consumer and computer applications [16, 17].

7.3.3.2 USB Interfaces

Universal serial bus (USB) is a standard interface between a computer and many peripheral devices, such as mouse, keyboard, printer, memory stick, mobile Internet modem, speaker, webcam, digital camera, external hard disc, smartphone, etc. [18]. Their number is increasing rapidly, because of its ease to connect and run more devices simultaneously. Modern laptops may have six or more USB ports.

The computer acts as a host and is also able to supply the current to most of the connected USB devices. There are three main generations of USB currently on the market, with different performance:

USB 1.0/1.1, with a data rate of 12 MB/s serving low- and full-speed applications, was the first USB generation and introduced in 1996.

USB 2.0, with a data rate of 480 MB/s serving high-speed applications, was introduced in 2002. This interface is specified to deliver a maximum current of

Fig. 7.4 Examples of USB adapters: (**a**) game port to USB adapter, (**b**) serial to USB adapter, (**c**) USB to Xbox adapter, (**d**) wireless USB adapter for iPod shuffle

500 mA at 5 V power supply. Applications that require larger currents must have their own power supply.

USB 3.0 with a data rate of 5 GB/s serving super high-speed applications was introduced in 2009. This generation has a connector with additional pins to achieve this extremely high speed, but it is still compatible with the other USB interface connectors.

There are also many adapters on the market, which allow the connection of conventional devices as well. Fig. 7.4 shows a couple of examples.

The fact that every standard USB device will work immediately upon being plugged into a computer, because it is automatically recognised and configured, is one of the main reasons for USB popularity. It is therefore expected that more and more electronic devices can be connected through a USB interface.

7.4   Mixed-Signal ICs and Integration

Due to the convergence of digital communications, consumer and computer, there is an increasing number of real-time signals to be processed: voice, professional audio, video, graphics, telephony, data streams, etc. This is usually performed by high-performance analog and digital signal processors.

Today's integrated circuits are complex heterogeneous systems: they combine complex analog, mixed-signal and RF capabilities with memory cores and digital cores such as processors, peripherals, codec and other complex functions, integrated on one single chip. Figure 7.5 shows a typical example of a heterogeneous chip in which the various cores are clearly visible.

The function of a chip is defined by the specification of its required behaviour. The complexity of its design is such that many of the applied digital, analog, memory and interface cores are not designed from scratch but are picked from an existing library of so-called *intellectual property* (*IP*) cores. This increases the design efficiency. This *reuse* not only requires easy portability of these cores between different ICs and between different technology nodes but also between different companies. Standardisation and quality of the IP cores are thus key requirements for an effective design reuse process.

Fig. 7.5 A digital audio broadcasting chip built from various cores

A broad, flexible and reliable library of pre-verified, standard compliant analog, digital, memory and interface cores is a minimum requirement to achieve first-time-right silicon with very short turnaround times. Therefore, many of these elements have been developed as a kind of (standard) IP. They may be available from an in-house library or may be licensed from an external company. Particularly programmable IP cores, such as microprocessor cores, come with specific application software.

Examples of IP are:

- Microprocessors (CPU): use software to control the rest of the system

    - Intel, SPARC, RISC-V, PowerPC, ARM Cortex, MIPS, 80C51, etc.

- Digital signal processors (DSPs): manipulate audio, video and data streams

    - Omaps, TMS320 and DaVinci (TI), DSP56000 series (Freescale), DSP16000 series (Agere), Oak, TeakLite.
    - Most DSPs are for wireless products.

- (F)PGA-based accelerators: decoders, encoders, error correction, encryption, graphics or other intensive tasks
- Memories

    - Synopsys, Artisan, ARM, embedded memories and caches
    - Memory controllers: controlling off-chip memories, Rambus, Synopsys, Altera

- Interfaces: external connections

    - USB, FireWire, Ethernet, PHY, UART, Bluetooth, keyboard, display or monitor

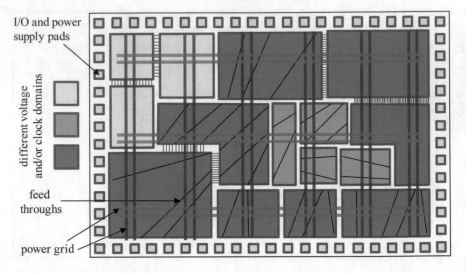

Fig. 7.6 Example floor plan of a chip

- Analog

  - A/D, D/A, PLL (e.g. for use in clock generation), oscillator, operational amplifier, differential amplifier, bandgap reference

So, a chip consists of a combination of reused (IP) cores and newly designed cores. Reuse is not only beneficial to reduce the amount of hardware development, but particularly in case of programmable cores, it certainly also reduces the amount of software development. Both the operating system software and much of the application software are readily available from various suppliers, when using standard (micro-)processor cores. These suppliers have concurrently developed the hardware and software for these *embedded architectures* [19].

When the design of the individual cores is completed, the various cores must be integrated on the chip. This is a complex job, since a chip may consist of tens to several hundreds of cores, each with their own layout and shape. This requires an optimum *floor plan*, which also deals with the intensity of data transfer between the cores and with power consumption aspects, as shown in Fig. 7.6.

After the chip design has been completed, it has to be extensively verified. This includes both a complete top-level functional and timing *simulation* and *verification*. Functional simulation and verification verifies if the chip performs according to the specified functionality. Timing simulation and verification verifies whether the correct function can also be guaranteed under specific chip process and environmental conditions. These conditions must be within certain standardised limits (e.g. acceptable temperature operating range for military products: −55°C to 125°C or supply voltage between 90% and 110% of the nominal voltage. Within these limits the chip performance (speed and power consumption) should be guaranteed.

Although there are many tools that support the final verification steps, the giant complexity, both in terms of number of transistors (hundreds of millions to several billion) and frequency (several gigahertz), makes verification one of the most time-consuming stages of the total design trajectory.

After full chip verification, the database in *gds-2* format is now almost ready for chip fabrication. Such a gds-2 description contains all physical layout details of all components and interconnections on the chip. Because the chip is built from different layers, the database contains the complete information of all individual items in every layer. Eventually, the complete pattern in each layer has to be copied onto individual masks that are used during the photolithographic steps during fabrication. So, a database contains the description of all patterns in every *mask*. Advanced integrated circuits require between 40 and 65 different masks. The final step, before mask making, is to subject the database to a final *chip-finishing* stage. It includes several *design for manufacturability* (*DfM*) steps to improve yield, after which it is prepared for the mask fabrication process. It adds the *scribe lane* around the edges of the chip (for cutting the wafer into dies), a seal ring between the chip's periphery and the scribe lane (to prevent possible cracks during the stress that occurs during the cutting process) and mask alignment markers.

Now, the chip is ready for production. The following chapters gradually introduce all aspects that are related to the production of a chip, starting with the substrate on which they are fabricated.

7.5   Conclusion

Chapters 5, 6 and 7 showed the large variety of circuits and IP that can be integrated onto a single chip. Analog, digital, RF, mixed-signal and memory cores perform the functional and storage operations on a chip, while the communication with other parts of the system is performed by the interfaces and I/Os.

The increasing complexity and performance of both the ICs and systems put stringent requirements on the bandwidth of the interfaces between the various components of a system. It has elevated the number of bond pads on a consumer or communication IC to even above 1000. This, in turn, had severe impact on the development of the packages in which these chips are mounted. Chapter 14 presents an overview and trends in package technologies.

The large number of interface bits, together with their increasing frequencies, made I/Os to become part of the biggest sources of noise, which could obstruct proper operation of the individual ICs and of the system as a whole. Therefore, a lot of attention must be given to both the design of the individual I/O cells and to the way in which so many of these cells can be used in the overall chip design while keeping the noise at acceptable levels.

The increasing analog and mixed-signal contents, the growing complexity in terms of functionality and performance of a chip and the large number of I/O pins have made IC design and its verification the most costly and time-consuming engineering task in the overall development of an integrated circuit. A detailed discussion on the cost involved in chip development can be found in Chap. 15.

Chapter 8
Wafers and Transistors

8.1 Introduction

Metal oxide semiconductor (MOS) devices and integrated circuits can only be realised with semiconductor materials. The *substrate material* on which they are built plays a very dominant role in their performance and reliability. Aware of this fact, Lilienfeld used copper sulphide as a semiconductor starting material in 1930. Germanium was used during the early 1950s. Until 1960, however, usable *MOS transistors* could not be manufactured. During the 1960s, the move from germanium to silicon was made in which the transistor channel is a thin conductive layer, which is realised electrically. Table 8.1 shows a comparison of a few germanium (Ge) and silicon (Si) material constants, which clearly shows a few reasons why silicon is used today. For example, the circuits designed in the 65 nm CMOS node use transistors with a channel length close to 60 nm while operating at 1.2 V. This causes an electric field in the channel of 20 V/μm, which is much more than the maximum breakdown field in germanium but still less than that in silicon. Another reason is the widely accepted military spec, which can be met with this material. This spec requires products to function correctly at a maximum operating temperature of 125 °C. The maximum operating temperature of germanium is only 70 °C, while that of silicon is 150 °C. A major reason for these maximum temperatures is that, for many applications, the leakage currents become unacceptably large above these temperatures. Another advantage of silicon is that it is cheap because it is widely available as (beach) sand.

The breakthrough for the fast development of MOS transistors came with advances in planar silicon technology and accompanying research into the physical phenomena at and below the semiconductor surface.

We will start this chapter with a discussion on the basic architecture of the *metal oxide semiconductor* (*MOS*) transistor. Also the differences between nMOS and pMOS transistors will be explained. They are the basic devices to create *complementary MOS* (*CMOS*) chips, which represent more than 90% of the total chip market. Bulk silicon has long been the starting wafer material for conventional (C)MOS

© Springer International Publishing AG, part of Springer Nature 2019

H. Veendrick, *Bits on Chips*, https://doi.org/10.1007/978-3-319-76096-4_8

Table 8.1 Important material constants from germanium and silicon

Material constant	Germanium	Silicon
Melting point [°C]	937	1415
Breakdown field [V/µm]	8	30
Relative expansion coeff. [°C]$^{-1}$	5.8×10^{-6}	2.5×10^{-6}
ε_r	16.8	11.7
Max. operating temp. [°C]	70	150

processes, while several high-performance microprocessors are made on SOI wafers. Some of the important characteristics associated with these different substrates (wafers) are also discussed.

8.2   What Is an n-Type and What Is a p-Type Semiconductor?

Until the mid-1980s, the nMOS silicon-gate process (see Sect. 10.3.1) was the most commonly used process for MOS LSI and VLSI circuits. Their average power consumption, however, had gradually increased to 1 W to 2 W at that time, which was close to the maximum power that could be handled by the cheap plastic packages. CMOS circuits consume ten to fifteen times less power than their nMOS counterparts, which is the reason why they dominate the semiconductor market since then. CMOS circuits use two complementary operating devices: an nMOS and a pMOS transistor, cross sections of which are shown in Fig. 8.5. To explain the creation of n-type and p-type silicon, we start with a short discussion on the basic silicon crystal structure.

The structure of a free *silicon atom* is shown in Fig. 8.1 (left). This atom comprises a nucleus, several inner shells and an outer shell. The nucleus contains 14 protons (positively charged atomic particles) and 14 neutrons, which are particles with no charge, while the shells contain 14 electrons. Ten of these electrons are in the inner shells and four are in the outer shell. The positive charge of the protons and the negative charge of the electrons compensate each other to produce an atom with a net neutral charge.

The electrons in the shells may possess certain energy levels. These energy levels are grouped into *energy bands*, which are separated by energy gaps. An energy gap represents impossible levels of electron energy. Figure 8.1 (right) shows these bands and the energy gap for a typical solid material. The four electrons in the outer shell of a silicon atom are in the material's valence band and determine the physical and chemical properties of a material.

Figure 8.2 shows the bonds that these four electrons form with four neighbouring atoms to yield a silicon crystal (Fig. 8.7). *Monocrystalline silicon* consists of a large volume (array) of repeating patterns of regular-positioned silicon atoms.

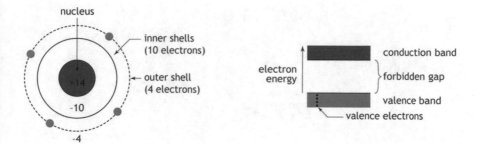

Fig. 8.1 Structure of a free silicon atom (left) and schematic representation of electron energy bands in a typical solid material (right)

Fig. 8.2 Silicon crystal

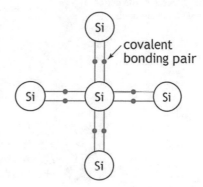

Because the conduction and valence bands in a conductor partly overlap, as shown in Fig. 8.3a, the electrons in a conductor can easily go from the *valence band* to the *conduction band*, which enables good conduction. In an insulator, however, none of the valence electrons can reach the conduction band. Figure 8.3b shows the large band gap generally associated with insulators. A semiconductor lies somewhere in between a conductor and an insulator. The associated small band gap is shown in Fig. 8.3c. Valence electrons may acquire sufficient energy to reach the conduction band, which produces a limited conduction mechanism in a semiconductor, compared to a conductor.

The most popular semiconductor substrate materials have four electrons in the outer shell and are therefore located in group IV of periodic system of elements (Table 8.2).

The introduction of an element from group III (with three electrons in the outer shell) or group V (with five electrons in the outer shell) in a semiconductor (group IV) crystal produces an *acceptor* or a *donor* atom, respectively. This process of replacing silicon atoms in a silicon wafer by atoms from another group is called *doping*. These doping atoms, also called *dopants*, take the position of a silicon atom. This semiconductor doping process dramatically changes the crystal properties.

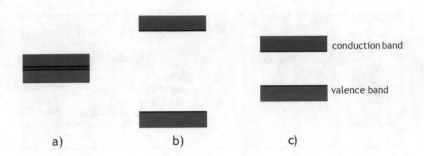

Fig. 8.3 (a) Energy bands of a conductor, (b) an insulator, (c) an intrinsic semiconductor

Table 8.2 Various materials as grouped in the periodic system of elements	Group		
	III (Acceptors)	IV	V (Donors)
	Boron	Carbon	Nitrogen
	Aluminium	Silicon	Phosphorous
	Gallium	Germanium	Arsenic
	Indium	Stannic (tin)	Antimony

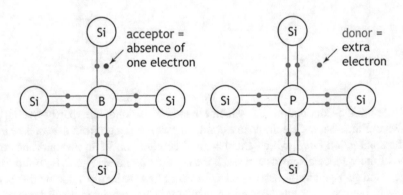

Fig. 8.4 p-type silicon, e.g. when silicon (group IV) is doped with boron (group III), and n-type silicon, e.g. when silicon (group V) is doped with phosphorous (group V)

The presence of a group III atom in a silicon crystal lattice is considered first. The situation for boron (B) is illustrated in Fig. 8.4 (left). Boron has one electron less than silicon (compare with Fig. 8.2), and it therefore lacks one electron required for a bond with one of its four neighbouring silicon atoms. The 'absent electron', which is called a *hole* in the resulting p-type semiconductor, is a willing *acceptor* for an electron from an alternative source. In fact, when a hole position accepts an electron, that electron leaves another hole behind, which can now accept another electron, etc. So, in such a *p-type semiconductor*, the current is caused by the movement of holes (positive charge).

Similar reasoning applies when a group V atom, such as phosphorus (P), is present in the silicon lattice. This situation is illustrated in Fig. 8.4 (right). The extra electron in the phosphorus atom cannot be accommodated in the regular bonding structure of the silicon lattice. It is therefore relatively easy to move this *donor* electron through the resulting *n-type semiconductor*.

Because an acceptor lacks an electron (a hole = lack of one electron), while a donor has one extra electron, the mobility of electrons is usually larger (about two to three times) than that of holes. In other words, an *nMOS transistor* (current caused by electrons) usually has a larger current and a higher performance than an equally sized *pMOS transistor* (current caused by holes).

The above theory concerning the different types of semiconductors will now be used to explain the architecture and behaviour of the MOS transistor.

8.3   Basic MOS Transistor Architectures

CMOS circuits operate with two complementary operating devices: an nMOS and a pMOS transistor. Figure 8.5 presents a cross section of both transistors to show their basic architectures on bulk silicon and explain their operation.

While nMOS transistors are fabricated on a p-type substrate, pMOS transistors use n-type substrates. The indication p^- refers to low p-type doped substrate, which means that only a few silicon atoms have been replaced by boron atoms, while indication n^+ refers to a heavily doped n-type source and drain regions, which means that in these regions many silicon atoms have been replaced by phosphorous or arsenic atoms.

A *MOS transistor* basically has four terminals: gate, source, drain and bulk (substrate). In the following, we will explain MOS transistor operation, starting with the nMOS transistor. The p^--substrate of this transistor is usually connected to ground (0 V). Between the gate and this substrate, there is a very thin oxide, called the *gate oxide*, with a thickness (t_{ox}) of only a few nanometres. In this way, the gate can very well control the conduction properties of the region in the top of the substrate right below the gate. This is called the *transistor channel* region. The transistor has a *channel width W* and a *channel length L*.

A MOS transistor has a built-in threshold before it can conduct. This is usually expressed as the *threshold voltage V_t*. An nMOS transistor has a relatively small positive V_t, usually between 0.2 and 0.5 V. Its value can be adjusted by the number of doping atoms in the transistor channel. As long as the gate voltage (with respect to the source) is set to zero (0 V, which is below V_t), there is no conduction possible between the n^+-source and drain (Fig. 8.5a). However, when we put a positive voltage on the gate with respect to the source, first the positive holes are forced away from the substrate region right below the gate. When the gate voltage exceeds the threshold voltage, then an n-channel is created in this region (Fig. 8.5b). This n-channel then connects the n^+-source and drain and allows charge carriers to flow from source to drain. In an n-channel device, these carriers are negatively charged

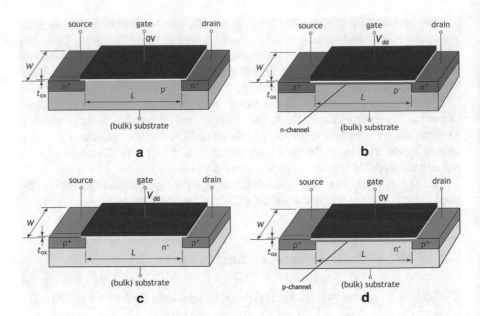

Fig. 8.5 Basic architecture of an nMOS transistor and a pMOS transistor: (**a**) non-conducting nMOSt, (**b**) conducting nMOSt, (**c**) non-conducting pMOSt, (**d**) conducting pMOSt

electrons. The charge carriers that flow through the device cause a current. The amount of current that flows through the channel depends on the gate voltage (= gate-to-source voltage), on the voltage difference across the channel (= drain-to-source voltage), on the gate-oxide thickness and on the width and length of the channel. The relation between these parameters and the transistor current is given in the expression in Sect. 5.1. The wider the channel, the more current can flow through it. The longer the channel, the more resistance between source and drain will be experienced, and the less current can flow through it. Usually the W/L ratio of the channel is used to define the current capability of the transistor. So, individual transistors can have different W/L ratios. Transistors with large W/L ratios are normally used in circuits that need to drive large load capacitances, like long on-chip buses, *clock drivers* (a driver is also called a buffer) or *output drivers* in the chip periphery that need to drive the off-chip (on-board) buses and the inputs of externally connected ICs.

When the gate voltage is set back to zero, the channel disappears, and source and drain regions are isolated from one another again so that the transistor no longer conducts.

Compared to the nMOS transistor, the pMOS transistor (Fig. 8.5c) is not only complimentary regarding its physical architecture (p⁺-source and drain regions instead of n⁺ regions and n⁻-substrate (bulk) instead of p⁻-substrate), it also operates complementary. Its n⁻-substrate is usually connected to the supply voltage (V_{dd}). A pMOS transistor has a negative V_t, meaning that it can only conduct when its gate is at least a threshold below the source voltage, assuming that the source is connected to V_{dd}. When the gate is also put to a high voltage (V_{dd}), the gate-to-source voltage is zero, and no current can flow (Fig. 8.5c). However, when the gate voltage is set to zero, it is far more than a threshold below the source voltage, and a p-channel is created in the region below the gate (Fig. 8.5d). This p-channel then connects the p^+-source and drain regions and allows charge carriers to flow from drain to source. In a p-channel device, these carriers are positively charged *holes*. A hole is in fact an empty electron position. It can capture an electron from another location, which then, on its turn, leaves another hole behind. So, in a pMOS transistor, the current is caused by a flow of holes, while in an nMOS transistor, it is caused by electrons. Also here, by changing the geometry (W/L ratio) of the pMOS transistor, we can define the amount of current through the transistor, depending on its application. The majority of CMOS processes are so-called *bulk CMOS processes* in which the transistors are fabricated in a silicon (n-type and/or p-type) substrate.

8.4   Different Substrates (Wafers) as Starting Material

As said before, the substrate plays a major role in the performance, reliability and yield of semiconductor chips. We will therefore present some background on the various substrates (wafers) on which these chips are built today.

8.4.1   Wafer Fabrication

For a description of silicon growth and wafer fabrication, the reader is advised to read Part I, Sect. 4.3.

8.4.2   Wafer Sizes

From an economical perspective, larger wafers have led to reduced IC manufacturing costs. They allow more chips to be simultaneously processed. This rule drove the *wafer diameter* from close to 1 inch ($\approx$25 mm) about four decades ago to 12 inches ($\approx$300 mm) today (see Part I, Fig. 4.3). This has put severe pressure on keeping the wafer flatness, its resistivity and its low crystal defect density homogeneous across a rapidly increasing wafer area. Next to this, the introduction of a new wafer diameter generation requires a huge amount of development costs. A possible move from 300 mm to 450 mm wafer fabs has been put on hold because it is questionable whether the huge investment could ever be earned back [20].

8.4.3   Bulk and Epitaxial Silicon Wafers

Bulk silicon wafers are monocrystalline silicon wafers as discussed in Sect. 8.2. *Epitaxial wafers* consist of a thin, monocrystalline silicon layer grown on the polished surface of a bulk silicon wafer [21]. This so-called *epi layer* is defined to meet the specific requirements of the devices in terms of performance, isolation and reliability. This layer must be free of surface imperfections to guarantee a low defect density and limit the number of device failures. Since the electrons in a MOS transistor channel only travel in the surface region of the device, the epi layer thickness can be very small and is only defined by the transistor architecture (source/drain and isolation depths). It ranges from one to a few microns. Today, the quality of wafers produced by the Czochralski process (Part I, Sect. 4.3) is so high that an additional epitaxial layer is no longer required. Usually the total *wafer thickness* is typically 750 µm to avoid warping during high-temperature processing steps but may range between 400 µm and 1 mm, depending on the wafer size and technology node. It means that the major part of the wafer mainly serves as a substrate carrier for the ICs made on it. Many advanced products, such as smartphones, tablets, smart cards and USB sticks, require the stacking of ultra-thin components. Some of these applications require wafer thinning to 75 µm or even less.

Although the resistance of this substrate hardly affects the performance of digital circuits, it has influence on the robustness of the ICs built on it. Most conventional CMOS processes, including the 180 nm node, use/used the epitaxial layer on *low-resistive* (heavily doped) *substrates*, in order to reduce the chance of causing a reliability problem called *latch-up*. Next to the creation of nMOS and pMOS transistors in a CMOS process, it also creates combinations of parasitic transistors, which are called thyristors. Once triggered, these thyristors can latch one or more circuits into a fixed state ('0' or '1'), causing operational failures and reliability problems. Circuits fabricated on low-resistive substrates are much more difficult to be put in a latch-up state than circuits made on high-resistive substrates. With reducing supply voltages, the chance for triggering a parasitic thyristor to initiate latch-up is also diminishing. This, combined with the increasing integration of GHz RF functions, has made the use of *high-resistive* (lightly doped) *substrates* very popular from the 120 nm CMOS node onwards. It leads to performance increase of passive components, such as inductors, but also to a better electrical isolation between the noisy digital circuits and the sensitive RF and analog ones. Because the full device operation occurs within the top of the wafer, the wafer fabrication must be completely homogeneous with an extremely low crystal defectivity.

Stand-alone memories and high-speed microprocessors may require their own substrates. This shows that not all ICs must be made on the same substrate. The following subjects discuss substrates that enhance the device performance.

8.4.4   Crystal Orientation of the Silicon Wafer

The intrinsic speed of an electronic circuit or transistor is, in first approximation, proportional to its current. On its turn, in a MOS transistor, the current is very much related to the ease with which the charge carriers can move through the transistor channel. This is usually referred to as the *carrier mobility factor*, or shortly the *mobility*. The carriers in nMOS transistors are electrons, while they are holes in pMOS transistors. The continuous scaling of the transistor sizes has had a dramatic effect on the mobility. Therefore, a lot of research has been and is still being performed in a variety of ways to improve this mobility and to increase the transistor current per unit area. In this respect also the *crystal orientation* of the silicon substrate plays an important role.

A crystal is defined by the properties of its unit cells and recognised by a certain repeating pattern of atoms, which is typical for a certain material. In the unit cell, the arrangement of its atoms is often represented using coordinates. Fig. 8.6a shows that a *silicon crystal* (and also a germanium crystal) has a diamond structure. Figure 8.6b shows the orthogonal coordinate system. Planes in a crystal can be specified using so-called Miller indices and are expressed as (*xyz*).

For example, a plane that does not intersect the *y*- or *z*-axis is called the (100) plane. A plane in *z*-direction that intersects the *x*- and *y*-axis at equal distance from the origin would be (110), while a diagonal plane that intersects the *x*-, *y*- and *z*-axis at equal distance from the origin would be the (111) plane. Because the atomic arrangement and atomic density of silicon are different for each plane, the physical, chemical and electrical properties depend on the wafer orientation. Figure 8.6c shows the atomic arrangement in the various planes. It shows that the (111) plane has more atoms/cm^2 than the (110) plane, which, on its turn, has more atoms than the (100) plane. The (111) plane, for instance, oxidises much faster than the (100) plane but etches much slower. Also the mobility of the charge carriers throughout the crystal is also different for each plane.

Traditionally, CMOS has been fabricated on wafers with a (100) crystalline orientation, mainly due to the high electron mobility and low interface trap density. However, the pMOS transistors on this substrate suffer from low hole mobility. By moving away from the (100) orientation, *electron mobility* is degraded, while *hole mobility* is improved. Compared to a traditional (100) wafer, a (110) wafer can show hole mobility improvements up to 30% in practice (leading to 30% faster pMOS transistors), while electron mobility may have a little degraded by about 5–10% (leading to 5–10% slower nMOS transistors). If the pMOS channel is oriented along the <100> direction on a (100) wafer, its mobility and performance may be increased by about 15%, with almost no degradation of the nMOSt performance. This is only a minor change in the starting wafer, with no further consequences for the process technology and layout (Fig. 8.8). Traditionally, the wafer *notch* is cut during crystal grinding in the <110> direction (Fig. 8.7a). To orient the channel direction along <100> only requires a crystal rotation of 45^0 to grind the notch in <100> direction

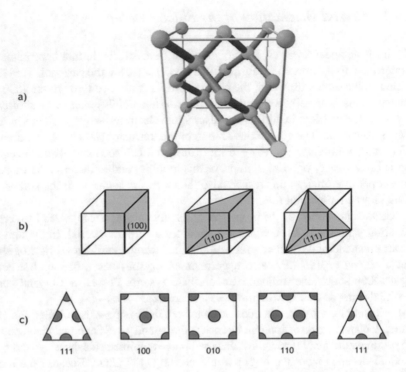

Fig. 8.6 Example of using Miller indices to represent crystal planes. (Detailed descriptions of the individual sub figures **a**), **b**) and **c**) can be found in the text)

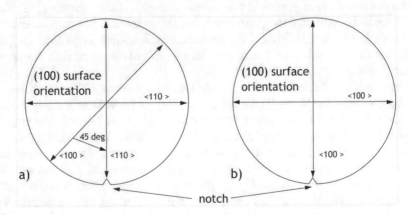

Fig. 8.7 (**a**) Traditional notch grinding and (**b**) grinding the notch in the <100> direction (Source: MEMC)

(Fig. 8.7b). This orientation change is a low-cost solution to enhance the pMOS device, logic gate and memory cell performance with no risk or consequences for the integration process. This wafer option is already in use in high volume production.

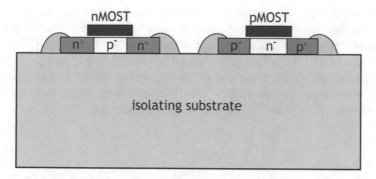

Fig. 8.8   Cross section of a basic SOI-CMOS process

8.4.5   Silicon-on-insulator (SOI)

Bulk CMOS devices, as shown in Fig. 8.5, show relatively large source/drain capacitances, because these are fabricated within the semiconductor's substrate. This can be avoided with the *silicon-on-insulator CMOS (SOI-CMOS)* devices illustrated in Fig. 8.8. In bulk CMOS processes, there is always the possibility for parasitic currents flowing through parasitic transistors, from the nMOS to the pMOS transistors or vice versa. This *latch-up* effect can put the circuit in a fixed but unrecoverable state. The complete isolation of nMOS and pMOS transistors associated with SOI-CMOS prevents the existence of such parasitic current paths and completely removes the latch-up reliability problem.

Because sources and drains are made on top of an isolating substrate, the parasitic source and drain capacitances are much less than those of the bulk CMOS processes. This makes the SOI-CMOS process particularly suitable for high-speed and/or low-power circuits. Sapphire was originally used as the isolating substrate in SOI circuits, despite the fact that it is substantially more expensive than silicon. Today, there are several cheap alternatives to produce SOI wafers: the SIMOX (separation by implantation of oxygen) [22] process and smart-cut process [23]. Further process details are beyond the scope of this book.

For advanced low-voltage CMOS (≤ 1 V) system-on-a-chip designs with digital, analog and RF parts, SOI is expected to offer a better performance than bulk CMOS technology [24, 25]. SOI circuits are said to deliver more speed at the same power consumption or to consume less power at the same speed. Furthermore, SOI realises better isolation between digital, analog and RF parts on the IC. Those circuits will therefore be less affected by substrate noise. Additionally an SOI transistor has lower parasitic capacitances and consequently exhibits a better RF performance. SOI devices are thermally insulated from the substrate by the buried-oxide layer. This leads to a substantial elevation of temperature (*self-heating*) within the SOI device, which consequently modifies the output current-voltage characteristics of the device, showing negative conductance.

For the 32 nm node, most semiconductor manufacturers still use bulk silicon wafers for their main CMOS process technology. The main reason is that these unprocessed wafers are two to three times cheaper than SOI wafers. Generally we can say that SOI adds 10–15% to the overall fabrication cost of an IC.

Many other alternative device and process options have been developed to extend CMOS technologies to even beyond 10 nm devices. A flavour of these technology options in both the devices and interconnects is presented in Sect. 10.4.

8.5   Conclusion

The numerous electronic applications require a large variety in functionality and performance of the integrated circuits from which they are built. This chapter has illustrated that different wafer materials are available, because all requirements cannot be served by just one type of substrate (wafer) only. The chapter also explained why most integrated circuits are being fabricated on silicon wafers. It also showed how the substrate must be processed to create nMOS and pMOS transistors, as they are the elementary devices in CMOS integrated circuits.

Finally, not only the substrate material but also its crystal orientation plays a dominant role in the overall reliability and performance of integrated circuits.

Chapter 9
Lithography

9.1 Lithography Basics

The production of an IC requires a translation of its specifications into a description of the layers from which it will be built. Usually, the patterns in all layers are represented in a layout. The generation of such a layout is usually done via an interactive graphics display for handcrafted layouts (certain analog circuits and/or some basic digital cells) or by means of synthesis, place-and-route and floor-planning tools, as discussed in Chap. 7. Figure 9.1 shows an example of a complex IC containing several functional blocks, many of which consist of a combination of handcrafted, synthesised and memory blocks.

A complete design is subjected to functional, electrical and layout design rule checks. When these checks prove satisfactory, then the complete layout is stored in a *database* (*gds-2 file*).

As will be explained in the next chapter, a chip is built from many different physical layers, which each requires several production steps. However, before executing a production step, the wafer must be masked to define the areas on it that must be affected by this step and the areas that must not. These areas are defined by a pattern in the corresponding mask layer. So, the final database contains the complete information of all individual geographical shapes in every mask layer.

A software programme (post-processor) is used to convert this database into a series of commands. These commands control an electron-beam pattern generator (EBPG) or a laser-beam pattern generator (LBPG), which creates an image of each layer on a photographic plate called a reticle. Such a *photo mask* or *reticle* contains a magnified copy of the corresponding layer in the database. Most photolithographic systems use reticles that are four times the size (=16 times the area) of the real chip (example 4:1 reticle in Part I, Fig. 4.4). The reticle pattern is thus demagnified as it passes through the projection optics during exposure, thereby forming an image of the reticle pattern on the surface of the wafer. Defects that would be critical on 1:1 reticles are now four times smaller after demagnification using 4:1 reticles. This optical reduction also reduces the influence of possible harmful particles that cross

Fig. 9.1 Example of a complex signal processor chip, containing several synthesised functional blocks. (Source: NXP Semiconductors)

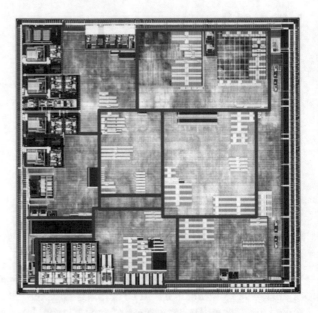

the light beam during exposure. Often pellicles are used to protect the reticle from these particles. A *pellicle* is a very thin transparent membrane adhered to a metal frame surrounding the reticle, with a standoff distance to the reticle of a few millimetres. It keeps particles out of focus during the lithographic process, so it will not image onto the wafer and reduces the possibility of printing defects. Figure 9.2 shows a layout of a reticle, including the alignment markers.

Small feature sizes, such as currently required in nanometre (<100 nm) CMOS processes, are obtained by using reduction scanners. The scanner exposes the pattern on the reticle repeatedly across the wafer in a grid. This is done by a zigzag movement of the wafer under the lens of the scanner. The 4:1 reduction is achieved by means of a very complex system of lenses. Figure 9.3 shows a basic schematic of a generic optical projection system.

The image reproduction quality of projection lithography tools is not only determined by this resolution. These tools also need to create clear, high-contrast images, while it must offer a sufficient *depth of focus (DOF)* to accommodate system focus variations and variations in height across the wafer.

The combination of a large number of metal layers and extremely large-area designs creates significant topographies across the individual chips and put stringent demands to the DOF. Therefore, many planarisation steps (Chap. 10) are used during chip fabrication to limit topology variations to below 40 nm.

Next to the fact that the *resolution* of the lithographic projections is limited by diffraction of the light, it also depends on the properties of the used photosensitive layer, called *photo resist*. Better photo resists allow smaller minimum feature sizes. To understand the limitations of and the alternatives for the current photolithography process, we need to understand the relation between the minimum *feature size*

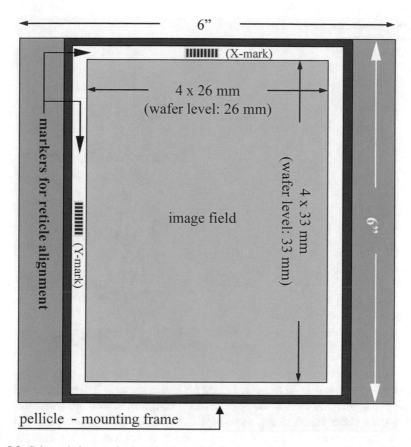

Fig. 9.2 Schematic layout of a 4x reduction reticle for step and scan systems. (Source: ASML)

on a chip and the wavelength of the light used during exposure. An expression, developed by Ernst Abbe around 1867, describes the feature size F, which usually refers to the smallest detail of the printed patterns:

$$F = \frac{k_1 . \lambda}{\mathrm{NA}} = \frac{k_1 . \lambda}{n \sin \alpha} \tag{9.1}$$

where k_1 is a constant, which is a function of the photo resist, the mask, illumination and *resolution enhancement techniques* (*RET*), which will be discussed later. With 'conventional' imaging, the value of k_1 was restricted to $k_1 \geq \frac{1}{2}$. With the use of RETs, k_1 can be further reduced to $\frac{1}{4} \leq k_1 \leq \frac{1}{2}$.

NA represents the numerical aperture of the lens, which characterises the range of angles (α in Figure 9.5) over which the lens can still capture an image from the object and n the refraction index of the medium between the lens and the wafer (1 for an air-based system).

Fig. 9.3 Basic schematic
of generic optical
projection system

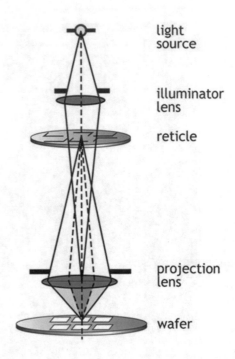

light
source

illuminator
lens

reticle

projection
lens

wafer

For many technology generations in the past, roughly until 350 nm processes, the values for k_1 and NA were close to each other, resulting in minimum feature sizes, which were about equal to the wavelength of the used light source, e.g. 0.35 μm feature sizes were mostly printed on i-line (365 nm) steppers. From a cost perspective, there is a strong drive to extend the use of the wavelength of the applied light source to smaller technologies. The 248 nm deep-UV (DUV) steppers, with a krypton-fluoride (KrF) light source, are even used for 90 nm feature sizes. The argon-fluoride (ArF) 193 nm DUV source could potentially be used for feature sizes until 60 nm with dry lithography. There are several resolution enhancement techniques that help to extend the use of 193 nm lithography to technologies with feature sizes even beyond 30 nm. These techniques are the subject of the next subsection.

9.2   Resolution Enhancement Techniques

Creating smaller feature sizes with the same wavelength requires compensation for non-ideal patterning, such as lens aberrations, variations in exposure dose, pattern sensitivity, die distribution across the reticle and the reticle size. The extension of the use of the 193 nm wavelength to sub 90 nm technologies would not have been possible without the use of several additional resolution enhancement techniques (RETs), such as optical proximity correction (OPC), off-axis illumination (OAI),

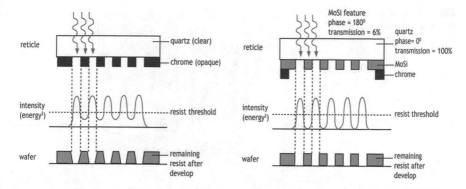

Fig. 9.4 Basic use of a binary photo mask (left) and an attenuated phase-shift mask (attPSM). (Source: ASML)

phase-shift masks (PSM), better resist technologies, immersion lithography and litho-friendly design. Although all of these techniques are currently applied, we will only discuss a few of them in some detail to present the reader a flavour of the increasing complexity and costs of the lithographic process, starting with the basic conventional binary mask.

Before the lithographic process starts, first a photosensitive layer, called *photo resist*, is deposited on the wafer. Then, with a conventional *binary mask*, in combination with the 193 nm light source, details as small as 90 nm can be depicted. A binary (photo) mask is composed of a quartz carrier with chrome features (Fig. 9.4 (left)) [26] that correspond to the designed layout pattern. Light passes through the clear quartz areas and is blocked by the opaque chrome areas. Where the light reaches the wafer, the photo resist is exposed, and these areas are later removed in the develop process, leaving the unexposed areas as features on the wafer. Binary masks are relatively cheap and they show long-life times, because they can be cleaned an almost infinite number of times. Moreover, they use the lowest exposure dose and enable high-throughput rates. Preferably all masks should be binary masks since it would reduce the overall chip production costs.

As on-chip feature sizes and spacings shrink, the resolution of the projection optics begins to limit the quality of the resist image. In the example in Fig. 9.4 (left), there is still significant light intensity below the opaque chrome areas, due to diffraction of the light through the very close neighbouring clear quartz areas. This 'unwanted' energy influences the quality of the resist profiles on the wafer after development. These profiles are ideally vertical to create accurate replicas of the mask patterns. Particularly the use of conventional binary masks with dense patterns of lines will produce a pattern of too narrow resist lines. When such masks are used in the definition of the metal interconnect layers, it will result in too narrow metal tracks on the chip, which may cause various electrical and reliability problems.

An alternative to the use of binary masks is the *phase-shift mask* (*PSM*) technology, which allows extending the limits of optical lithography. PSM technology is divided into two categories: attenuated PSM (attPSM) and alternating PSM

Fig. 9.5 Basic principle of
immersion lithography.
(Source: ASML)

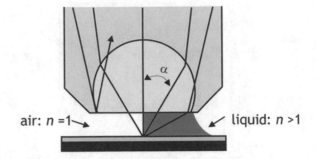

Fig. 9.5 Basic principle of
immersion lithography.
(Source: ASML)

(altPSM). To get an impression of the capabilities of the PSM technique, we will only discuss attPSM in some detail. In this technique, the resist image is formed by light through the patterns in the molybdenum silicide (MoSi) layer on the reticle (Fig. 9.4 right). Unlike chrome, MoSi allows a small percentage of the light to pass through (typically 6% or 18%) [26]. However, the thickness of the MoSi is chosen so that its transmitted light is 180° out of phase with the light that passes through the neighbouring clear quartz areas. The light that passes through the purple MoSi areas is too weak to expose the resist, and its 180° phase shift compensates the light intensity from the clear areas diffracted into these areas such that they appear to be 'darker' than similar features in chrome (Fig. 9.4 left). The result is a sharper intensity profile, which allows smaller features to be printed on the wafer. The 180° phase shift is only achieved for light at a given fixed wavelength. AttPSM masks can therefore only be used for one type of scanners only, while binary masks can be used for scanners with different wavelengths.

As explained, the above-presented lithographic techniques are basically applied to increase the resolution and/or depth of focus of the total illumination system. Another technique, which is already applied in several process nodes to enhance the lithographic properties, is called *immersion lithography*. When the photolithographic process is immersed in water ($n = 1.3$) and if we assume that $\sin\alpha$ in expression (9.1) can reach a maximum value of 0.95, then this 'water immersion lithography' can yield an *NA* close to 1.37. Only the lower part of the optics is immersed in water (Fig. 9.5). The left half in the figure shows the diffraction of the light beams in air, with a diffraction index $n = 1$ and some of the beams being reflected. The right half uses an immersion liquid with $n > 1$, which reduces the amount of reflected light, increasing the resolving power and allowing finer feature sizes. Immersion lithography also improves the *DOF*, which may resolve some of the related topography problems.

Using one of the above-described resolution enhancement techniques (RETs) is a prerequisite to create lithographic images with a satisfactory resolution and depth of focus. But it is not sufficient. When printing patterns with sub-wavelength resolution, they need to be compensated for the aberrations in the patterning. In other words, the fabricated IC patterns are no longer accurate replica of the originally designed patterns. So, we need already to compensate for these shortcomings by

Fig. 9.6 OPC using
SRAFs applied in the
mask-definition process.
(Source: ASML)

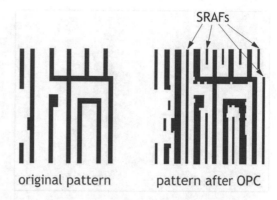

SRAFs

original pattern   pattern after OPC

changing the original mask patterns. Figure 9.6 shows how *optical proximity correction* (*OPC*) is applied in the mask-definition process. The right mask pattern is used during lithography to get the left (original layout) pattern image on the chip. Better imaging results can be achieved by using so-called sub-resolution assist features (*SRAFs*), such as scattering bars and hammerheads, which are too small to be printed onto the wafer but help to better reproduce the original layout pattern into the photo-resist layer.

This will certainly make the design process more complex. While the above-described RETs improve the resolution of the imaging system, the use of OPC masks will make them work. Mask costs, however, depend very much on the applied mask technology. When normalising the costs of a binary mask to 1, then an attPSM (without OPC) mask would cost 1.2 times as much and an attPSM (with OPC) mask 2.5 times. The use of altPSM is much more costly (six times more for altPSM without OPC and ten times more with OPC), since it requires an additional binary trim mask and thus needs double exposure.

Over time, many other innovations enabled extended use of photolithography beyond the 32 nm node. Also the support from the design side alleviates some of the problems when extending the use of 193 nm lithography further. To improve yield, complex *design for manufacturability* (*DfM*) design rules have already been used in many technology nodes. For the 65 nm node and beyond, this was certainly not enough. It also required strict *design for lithography* (*DfL*) design rules. DfL, also called *litho-friendly design*, is focused towards more regular unidirectional layout structures. It will simplify the lithographic process, it supports SRAFs, and it reduces mask costs. It may also lead to yield improvement due to a smaller variety of patterns to be printed.

Figure 9.7 shows two layout versions of a standard cell: the original layout with a plot of simulated line widths (a) and the litho-friendly layout with the corresponding plot of simulated line widths (b), showing more regularity. The red vertical lines in the respective layouts represent the polysilicon gates of the transistors. The widths of these polysilicon tracks determine the channel lengths of the transistors. This channel length is a very dominant parameter in the performance and leakage of a transistor and must therefore be very accurately reproduced. The simulated line

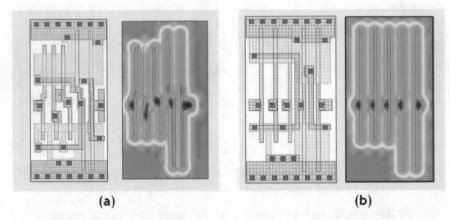

(a)   **(b)**

Fig. 9.7 Comparison of an original (**a**) and a litho-friendly layout (**b**) with more regularity. (Source: NXP Semiconductors)

widths of the litho-friendly regular pattern in figure (b) shows a much more homogeneous line width (channel length) of the polysilicon tracks than those in figure (a).

Next to the already discussed implications of RET and DfL for layout design, these techniques get more and more support from *EDA* (electronic design automation) tools and tool vendors and have become a standard part of the design flow.

Let's summarise the individual contributions of the above-described RETs: the combination of PSM, OPC and litho-friendly design may lead to a minimum k_1 to below 0.25, while water immersion can lead to a maximum *NA* of approximately 1.37. Substituting these values for 193 nm lithography in expression (9.1) leads to a minimum feature size of the smallest details of: $F \approx 30$ nm. Smaller line widths, e.g. 28 nm, can be obtained but with larger spacings.

The lifetime of 193 nm optical lithography is extended further by other advanced 'tricks'. Potentially, the feature size can be smaller by using a light-transparent liquid with a larger refraction index n than water. It is clear that all additional lithographic solutions to enable smaller feature sizes will reduce pattern aberrations, but not prevent them. According to today's knowledge, 193 nm lithography allows immersion liquids with a larger refraction index than that of suitable liquids in 157 nm immersion lithography. This latter lithography was expected to only extend the lifetime of photolithography for one more technology node. Industry has therefore decided to skip the 157 nm lithography because the development costs could not be recovered. Below 157 nm optically transparent materials used for lens and mask will become increasingly absorbent. A potential workaround for this absorption problem would be to build a completely *reflective optical system* based on mirror lenses. However, even by using the best-polished optical mirror devices, the reflection efficiency of the total optical system is limited to only a few percent. Extreme-UV lithography, which will be discussed below, also uses reflective optics.

Finally, particularly the semiconductor memory vendors have found a way to increase bit density without the use of very advanced and extremely expensive lithography tools. By using multiple layers of silicon (*3-D stacked silicon*), memory capacity can be increased dramatically, without increasing the footprint of the memory chip. Many flash memories are currently being fabricated using up to 48 stacked layers of memory cells. One-time programmable memories (OTPs) are also built from multiple layers already. NAND-flashes with 96 stacked memory cell layers are in development. *3-D technologies* are currently only economically viable when the complexity of the devices in these stacked layers is very limited. Because non-volatile memories use only one type of transistor in the cells (see Chap. 2) they are particularly suited for 3-D stacking. Therefore these additional layers are only used to fabricate arrays of memory cells. Many of the lithography steps, for patterning these layers, are replaced by smart etching steps, such that only one lithography step is needed for every five or six memory cell layers. Several of these stacked memory arrays have their peripheral address selection and sense amplifier circuits directly positioned in the wafer, while the memory cells are stacked on top of that.

9.3 Lithographic Alternatives Beyond 30 nm

Beyond 30 nm there are several alternative solutions. Only a few of them will be discussed here, just to illustrate the complexity of future lithography.

Use of an Immersion Liquid With higher refraction index n (see Eq. 9.1), creating a so-called superfluid NA in the range of 1.55–1.75. This would also require different lenses.

Use of a Multi-patterning Technology (MPT) When the pitch of two lines in a dense pattern is less than about 80 nm, it becomes a sub-resolution pitch, which can no longer be imaged accurately with 193 nm lithographic tools. Therefore this can be done with an image split: first image the odd lines with twice the minimum feature pitch and then image the even lines, also with twice the pitch (Fig. 9.8). This is usually referred to as *LELE* (litho-etch-litho-etch) *double patterning*.

A combination of double patterning and immersion lithography enabled the extension to or even beyond the 22 nm node. It allows lithographers to work with a higher k_1 (around 0.35 instead of 0.25). Although this eases the lithography process, DPT is more complex than a single-exposure process because it requires two masks and two exposures. The biggest challenge is the high accuracy of the alignment of the two masks during exposure. Also triple and quadruple patterning techniques have already been used for 16 nm CMOS processes and beyond. These techniques also contribute to increasing mask and processing cost. All multi-patterning techniques require an intelligent split of a single mask pattern into more separated masks, each with a lower-resolution pattern than the original pattern. In standard-cell design, this can be handled by tools; however, in optimised memory and analog circuit design, the designer faces additional design rules to fulfil the requirements of double (triple or quadruple) patterning. An example of the decomposition of an original layout

Fig. 9.8 Example of
LELE double patterning

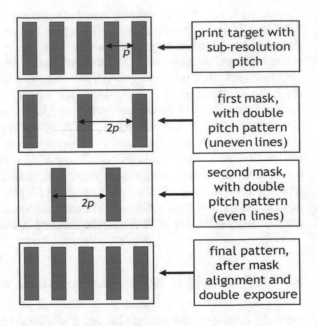

into two or three masks is shown in Fig. 9.9 [27]. This type of pattern decomposition
was used for process nodes between 22 nm and 14 nm. Actually, LELE lithography
has never become very popular. It causes severe problems with overlay and requires
doubling of the number of masks and exposures or more in case of
multi-patterning.

An alternative to LELE double patterning is the *self-aligned double patterning*
(*SADP*) and *self-aligned quadruple patterning* (*SAQP*). Figure 9.10 shows the basic
steps in SADP. In this technology the final pattern on the wafer is created by the
formation of sub-resolution features during semiconductor process steps, rather
than by sub-resolution lithography. The print target is shown in Fig. 9.10 (a). As a
first step, a hard mask layer is deposited or grown on the wafer. To support the formation of sub-resolution spacers, a sacrificial polysilicon layer is deposited on the
wafer and patterned with a relatively large optical lithography pitch (b). Since many
of the layers are deposited with an atomic layer deposition (ALD) step, where no
high temperature step is involved, the polysilicon is often replaced by photo-resist
material. Next, an oxide (or nitride or other) layer is deposited on top of the structure and then etched back until sub-resolution sidewall spacers are left (c). Then the
sacrificial polysilicon is removed (etched) (d), followed by a pattern transfer from
spacer to hard mask (e). Finally the pattern in the hard mask is used to create the
final pattern on the wafer (f). This spacer technology is a convenient approach to
achieve sub-resolution patterning with relatively large optical resolution pitches,
avoiding problems of, e.g. overlay between successive exposures in a double patterning technology.

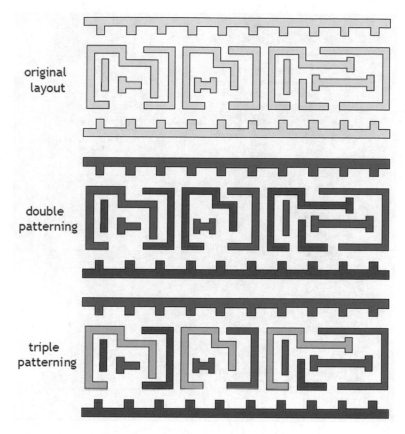

Fig. 9.9 Decomposition of an original layout into two or three individual masks. (Image: David Abercrombie; Mentor Graphics Corp)

SADP double patterning is often used in advanced memories, because memories typically consist of uniform pattern distributions. Currently spacer lithography is also increasingly used in the formation of the fins in FinFET process nodes of 20 nm and beyond. Even logic circuits in advanced FinFET processes are increasingly built from fully regular layout patterns in the creation of fins and transistor gates. Let us assume that we now use the pattern structure in step e in Fig. 9.10 as a starting point for a second SADP iteration and we repeat steps c to f, then we have again doubled the number of features. This is often referred to as self-aligned *quadruple patterning* (SAQP).

Use of extreme-UV (EUV) lithography With a light source wavelength of 13.5 nm, EUV was regarded as the most probable potential lithography solution for technology nodes beyond 28 nm. However, EUV 'light' is absorbed by all materials, including air. Therefore mirror lenses have to be used in a vacuum-based system with reflective instead of refractive optics and reticles. Over the last decade, a lot of problems have been solved before it could be used in high-volume production. A

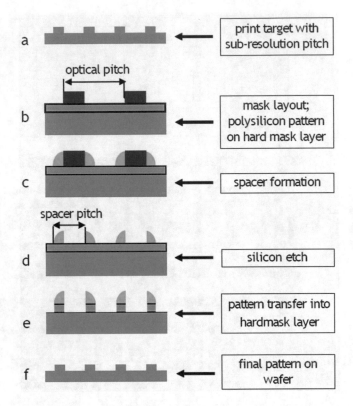

Fig. 9.10 Basic steps in spacer (SADP) lithography

few of them will be mentioned here. First, there was no suitable resist for high-volume production available. Second, because the light needs to propagate through a system with a large number of mirrors (nine or ten), where each mirror absorbs about 62% of the light intensity at the EUV-wavelength, extremely powerful light sources are needed in combination with relatively long exposure times. Figure 9.11 shows the light transmission path in an EUV scanner. It clearly shows the number of reflections which contribute to the loss of light intensity.

To generate 1 W of EUV power, the RF power needed to activate the plasma light source (red dot in the figure) may be as high as several hundred kilowatts. The lack of appropriate power sources and resists was a major bottleneck for bringing this lithography to the market. This explains the need for an improved light reflection of collector, lenses and reticle to improve the throughput time and reduce the power consumption. In 2006, the first EUV lithography tools (demo tool, US$ 65 million!!) have already been shipped. It was not meant for production, but it will support R&D programmes at IMEC (Leuven, Belgium) and at CNSE (University of Albany, New York).

Although current immersion scanners show throughputs of 175–275 wafers per hour, the effective throughput with double, triple or quadruple patterning reduces with a factor of two, three or four, respectively. Currently (2018) EUV wafer

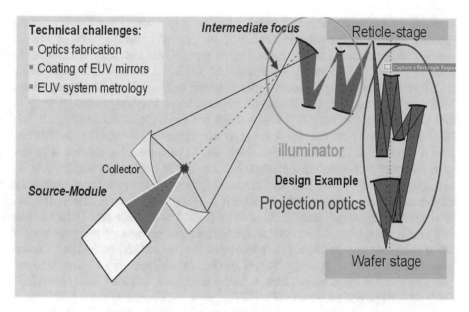

Fig. 9.11 The transmission path of the light in an EUV scanner as it travels from the light source to the wafer. (Courtesy of Carl Zeiss)

throughput is between 1000 and 2000 wafers per day, based on 80 W IF power. With 250 W IF power and improved resist sensitivity, this number could increase to around 125 wafers per hour, which makes EUV [28] very competitive with alternative multi-patterning lithography technologies for sub-10 nm nodes.

Use of Alternative Techniques to Fabricate Image-critical Patterns in Sub 30 nm Technologies For many years, *X-ray lithography (XRL)* has been a potential candidate for next-generation lithography (NGL). It uses X-rays, which generate photons with a wavelength of roughly 1 nm to expose the resist film on the wafer, enabling much finer features than currently with optical lithography tools. However, it has some major disadvantages. Conventional lenses are unable to focus X-rays, and, consequently, XRL tools cannot use a lens to shrink the mask features. Therefore its 1:1 pattern transfer methodology requires mask patterns with only one-fourth of the feature sizes used in the 4:1 photolithography masks. In addition, it requires an extremely expensive synchrotron, which converts an electron beam into an X-ray beam. It is therefore expected that the use of XRL will be limited to fabrication processes that create niche devices such as MEMS.

An alternative to photolithography beyond the 28 nm node is the *nano-imprint lithography (NIL)*. This 1:1 technology is based on physically pressing a hard mould (typically identical to the quartz/chrome material commonly used for optical lithography) with a pattern of nano-structures onto a thin blanket of thermal plastic-resist layer on the sample substrate. This imprinting step is usually done with the resist heated, such that it becomes liquid and can easily be deformed by the pattern on the mould. After cooling down, the mould is separated from the sample, which now contains a copy of the original pattern. Its major advantage is that it can replicate features with nanometre dimensions [29]. This process is already used in volume production in electrical, optical and biological applications. For semiconductor applications, the *step-and-flash* imprint lithography (*SFIL*) seems to be the most viable one. It allows imprinting at room temperature with only a little pressure using a low-viscosity UV curing solution instead of the resist layer. The higher the sensitivity to UV, the less exposure time the solution needs and the higher the throughput. In this imprint technology, some of the wafer process complexity has moved to the fabrication of the mould. Still a lot of key issues need to be solved, but the results of this disruptive technology, so far, are promising. A potential barrier for using the imprint lithography is that it requires very advanced lithographic processes to create the nanometre patterns on the mould. Because it is a 1:1 pattern transfer process, the pattern dimensions are only one-fourth of those printed on a photo mask, which is one of its major challenges. Moreover, low throughput may become the real showstopper for this technology. Reference [29] discusses the process and potentials of nano-imprint in more detail. Recently, NIL is also seen as an alternative to photolithography in photonics applications, such as in the fabrication of LEDs and photovoltaic (PV) cells. For the fabrication of ICs, NIL has regained interest by a 3-D NAND flash manufacturer, as to reduce the production cost of NAND flash memories [30].

Line widths down to 15 nm are claimed, while the cost could be less than the use of quad-patterning techniques or EUV. The mould can be made using e-beam lithography.

Moore's law is driven by the economical requirements of the semiconductor markets. This means that all semiconductor disciplines (design, lithography, fabrication, packaging, testing, etc.) are cost driven. For the lithography it means that there is a constant drive to make masks cheaper or to use cheaper masks for certain low-resolution process steps. Binary masks are relatively simple and cheap, guarantee high throughput and can be non-destructively cleaned. As explained before, the patterns in the layers with the smallest feature sizes require complex masks, such as attenuated PSM masks and/or mask correction techniques such as optical proximity correction. AttPSM masks must be cleaned in a chemical liquid, which is a destructive process, such that they can only be cleaned about three times and are therefore much more expensive.

To minimise mask costs during the fabrication process, the expensive mask types are only used to image the patterns in those layers that really need the smallest feature sizes. For the production of one type of memory, for example, different mask categories can be used. For the production of a flash memory with 22 masks, for instance, 4 ArF (attPSM + OPC) masks, 12 KrF (6 binary and 6 attPSM) and 6 I-line (binary) masks may be used.

Finally, particularly the semiconductor memory vendors have found ways to increase bit density without the use of very advanced and expensive lithography tools (see Part II, Chap. 6). By using multiple layers of silicon (3-D stacked silicon), memory capacity can be increased dramatically, without increasing the footprint of the memory chip.

9.4   Mask Cost Reduction Techniques for Low-Volume Production

To share the masks costs for low-volume products or test chips, so-called multi-project wafers (*MPW*) are used. These are normal wafers, but they contain products from different companies and/or (research) institutes. Their designs are included on the same mask set to reduce overall mask costs (Fig. 9.12).

Another way to share the mask cost is the *multilayer reticle (MLR)*, on which several mask layers of the same product are grouped next to one another on a single mask to reduce the physical number of reticles. These MLRs do not combine designs of different products. Both techniques are particularly used for small-volume designs, for prototyping and for educational purposes. To save mask costs completely, *direct-writing techniques* use an *electron-beam (e-beam)* or *laser-beam* system, which writes the layout pattern directly onto a wafer-resist layer, without using a mask. It requires the deposition of an additional conductive layer on the resist layer, to prevent damage by the beam during the patterning process. The resolution yielded by an e-beam machine is better than 5 nm but at a much lower throughput because it writes every feature individually. It is free of wavelength aberration. Laser-beam systems are gaining market share at the cost of *e-beam systems*, because they are cheaper as they do not require a vacuum environment. Because of their low throughput, the use of e-beam and laser-beam, today, is limited to fabricate low-volume samples, such as MPWs, prototyping products and test silicon for process development. Next to that these techniques are used to fabricate the physical glass-chrome masks (reticles) for use in photolithography processes.

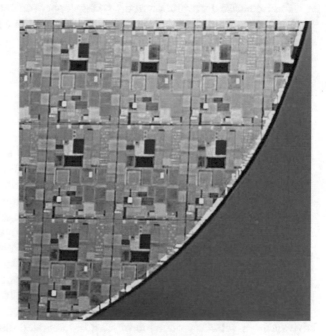

Fig. 9.12 Example of a multi-project wafer (MPW)

Table 9.1 The evolution of the wafer scanner. Source: ASML (2016)

Status @ 2016 of most advanced litho-tools						
Name		I-line	DUV	193	193i	EUV
Illumination source		Hg lamp	KrF laser	ArF laser	ArF laser	LPP
Reduction		4×	4×	4×	4×	4×
Wavelength	nm	365	248	193	193	13.5
NA$_{max}$ projection lens		0.65	0.93	0.93	1.35	0.33
k_{1_min}		0.6	0.3	0.3	0.3	0.4
Minimum pitch	nm	350	80	65	38	16
Overlay control	DCO (nm)	35	3.5	3.5	1.6	1.5
	MMO (nm)	60	5	5	2.5	2.5
Wafer size	Inch	8″/12″	8″/12″	8″/12″	12″	12″
Throughput	Wph	–/220	–/220	–/205	275	125

These direct-writing techniques are also called *mask-less lithography* (*MLL* or *ML2*) and are currently also being explored as an alternative for, or successor of the conventional photolithography, even for high-volume production. The main reason is the rapidly increasing costs of an optical mask set, which has reached the $2 million mark for the 45 nm node, although these costs will reduce due to the mask production learning curve. Over the last decade, a lot of progress has already been made to improve throughput. More information on future lithography techniques can be found in numerous publications and also on the Internet and is beyond the scope of this book.

To summarise the evolution of the wafer stepper/scanner, Table 9.1 presents several key parameters, which reflect the improvements made over different generations of steppers/scanners.

9.5 Conclusion

Photolithography is a critical step in the overall creation and cost of integrated circuits. The minimum sizes of the patterns on a chip have become so small that every time we scale them further, the photolithography process and equipment have to be adapted to be able to image the original layout design patterns with sufficient accuracy onto the dies on the wafer. It is clear that the capabilities of photolithography have been used already up to or close their limits. The use of the current 193 nm lithography has been extended to below 12 nm technologies, but that requires very advanced and costly additional techniques, multi-patterning and advanced process steps. In the meantime, alternative techniques such as extreme UV (EUV) or nano-imprint are further being developed and improved. Particularly EUV is already capable of imaging much smaller details than achievable with photolithography. With improved throughput it has currently started to become commercially attractive for high-volume production of CMOS-integrated circuits close to or beyond 10 nm.

Chapter 10
Fabrication

10.1 Introduction

Advanced nanometre CMOS processes, with channel lengths below 100 nm, have emerged from the numerous manufacturing processes that have evolved since the introduction of the MOS transistor in integrated circuits during the 1970s. Often the process-related discussions are either focussed towards that part of the process in which the transistors are fabricated or towards the part in which the contacts and interconnections are fabricated. These parts are called *front-end process* and *back-end process*, respectively.

Differences between processes were mainly characterised by the following features:

- The minimum feature sizes (e.g. transistor channel length and width, contact sizes, metal track widths and spacings, etc.).
- The gate-oxide thickness. This currently varies from 5 nm to 1 nm. The latter value resembles a thickness of only four to five atom layers!
- The number of interconnection levels. In most modern CMOS processes, this number is between 5 and 10 layers, depending on the application.
- The type of substrate material. Alternatives include n-type and p-type, high-resistive (e.g. p⁻) and low-resistive (e.g. p⁺) bulk silicon and epitaxial or SOI wafers, as discussed in Chap. 8.
- The choice of the gate material. Initially, the gate material was a metal implied in the acronym MOS (metal oxide semiconductor). Molybdenum and aluminium gates have also been used during the 1960s and 1970s. After that, almost all CMOS processes until the 90 nm node use polycrystalline silicon (*polysilicon*) as gate material because it perfectly interacts with the underlying silicon dioxide that was used as gate dielectric. From 90 nm onwards, a stack of W-WN-polysilicon and SiO_xN_y is used. A combination of a metal gate and high-k dielectrics is first introduced in the 45 nm node [31] but is more intensively used in the 32 nm node and below.

© Springer International Publishing AG, part of Springer Nature 2019
H. Veendrick, *Bits on Chips*, https://doi.org/10.1007/978-3-319-76096-4_10

- The method to isolate transistors. Today's processes use shallow-trench isolation (STI) (see this chapter).
- The type of transistors used: nMOS, pMOS, standard-V_t, high-V_t, low-V_t, planar or FinFET.

Modern fabrication processes consist of numerous photolithographic, deposition, etching, oxidation, implantation, diffusion and planarisation steps. These steps are frequently repeated throughout the process, and they currently total more than 1000 for the creation of an advanced chip. Most processes use between 25 and 70 masks, which together contain all the patterns that define the millions of transistors and interconnections from which a chip is built. The mask patterns are copied onto the dies on a wafer using lithography, the basics of which have extensively been discussed in Chap. 9.

In this chapter we will now gradually build the overall fabrication picture. Every process step is preceded by a sequence of pattern-imaging steps in which the mask pattern is copied onto the wafer. This is followed by a discussion on the basic process steps that are repeatedly used to create and define the various layers from which a chip is built: deposition, etching, oxidation, doping and planarisation. Then a simple process flow of a basic nMOS process with just five masks is presented by using transistor cross sections that correspond to the various phases of the process. This illustrates the relation between the various layers that together implement the transistors and interconnections on a chip. Subsequently, a basic CMOS process flow is briefly examined. Fundamental differences between various CMOS processes are then highlighted.

Next, an advanced nanometre CMOS process is summarised. Many of the associated additional processing steps are an extension of those in the basic CMOS process flow. Therefore, only the most fundamental deviations from the conventional steps are explained. Finally, several alternatives are discussed for both the transistors and interconnections to continue Moore's law to beyond 10 nm technologies.

10.2   Basic Process Steps

As mentioned before, the basics of photolithography are discussed in Chap. 9.

The sequence of copying a mask pattern onto the wafer is already discussed in Part I, Sect. 4.4. Once the mask pattern is copied onto the wafer, it is followed by a physical process step. Although the complete manufacture of a chip consists of up to a thousand individual process steps, these are often a repetition a just a few basic steps: deposition, etching, oxidation, implantation or diffusion and planarisation. To better understand the CMOS fabrication process, these steps will be described at an introductory level in the following sections.

10.2.1 Deposition

The deposition of thin layers of isolating material (dielectric), gate material (poly-silicon) and interconnections (metal) is an important aspect of the creation of transistors and interconnections on a chip. Sometimes, high temperatures (even more than 1000 °C) have to be used for the deposition. The growth of an *epitaxial film* (thin layer) (Sect. 8.4.3) is the result of a deposition step combined with a chemical reaction between the deposited and substrate material.

Metal layers are deposited by both physical and chemical methods. With physical methods, such as *evaporation* and *sputtering*, the material is physically moved onto the wafer. An example of this is the sputtering of aluminium. In this process an aluminium target is bombarded with argon ions, which physically dislodge aluminium molecules, which then flow from the target to the wafer surface. After deposition of the aluminium, photolithographic and etching steps are used to define the required metal pattern in it.

It was already known for more than a century that *copper* is a much better conductor than aluminium. However, copper cannot be deposited and etched as easily as aluminium. Copper diffuses through oxides and may reach the transistors, leading to changes in their operation and reliability. Process technologies with a copper back-end use a so-called *damascene process flow*, which is based on a metal inlay process. First, trenches are etched in an isolation layer. To prevent copper diffusion, a thin barrier layer is deposited, followed by the electroplate-deposition process of the copper, to improve copper adhesion and coverage copper. Copper tracks are then

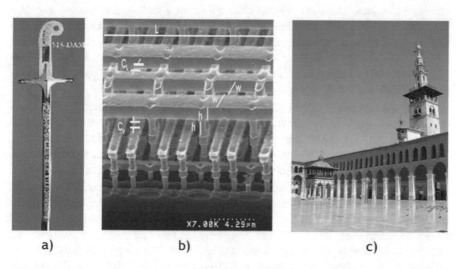

a) b) c)

Fig. 10.1 (**a**) The metal inlay technique was pioneered in Damascus in ancient times where it was already used in swords; (**b**) cross section of a chip, where all inter-metal dielectrics have been etched away to clearly show all contacts (vias) and metal layers, which resembles the ancient Arabic building style as used in the Umayyad Mosque in Damascus (**c**). (Photos: NXP Semiconductors and the Internet)

remaining as a metal inlay in the trenches (damascene processing), similar to the metal inlay in swords, made in ancient times in Damascus, Syria (Fig. 10.1a). The result of this processing is shown in Fig. 10.1b, which also compares with the ancient damascene building style as used in the Umayyad Mosque in Damascus (Fig. 10.1c).

Generally, the choice of deposition method is determined by a number of factors. One deposition step that got a lot of attention over the last decade is the so-called *atomic-layer deposition (ALD)*, particularly for its potential applications in advanced high-k gate dielectrics, in stacked Gate GAA NAND flash cells, in DRAM capacitor dielectrics and in copper diffusion barriers in advanced CMOS and memory processes. Without going deep into the chemical and physical reactions, ALD basically uses pulses of gas, creating one atomic layer at a time. So, the deposited film thickness is only dependent on the number of deposition cycles providing extremely high uniformity and accurate thickness control. It is used to create layers with a thickness close to 1 nm. It is therefore also of interest in all (sub-) nanometre applications that benefit from accurate control of (ultra-) thin films.

10.2.2   Etching

The previously described photolithographic steps produce a pattern in a nitride or equivalent barrier layer. This patterned nitride layer is also called a *hard mask*. It acts as a protection when its image is duplicated on its underlying layer by means of etching processes. There are several different *etching* techniques. The etching process must fulfil the following requirements: a high degree of anisotropy (unidirectional), good dimensional control, a high etching rate to minimise processing time, a high selectivity for different materials, a perfect homogeneity and reproducibility (e.g. several billion trenches in a 4 Gb DRAM) and a limited damage or contamination to satisfy reliability standards.

There are two major etching techniques: wet etching and dry etching. With *wet etching*, the wafer is immersed in a chemical etching liquid. The wet-etching methods are isotropic, i.e. the etching rate is the same in all directions (*isotropic etching*). The associated 'under-etch' problem illustrated in Fig. 10.2a became serious when the minimum line width of the etched layer approached its thickness.

Dry-etching methods may consist of both physical and chemical processes (anisotropic etching) or of a chemical process only (isotropic). Dry-etching methods, which use plasma, allow *anisotropic etching*, i.e. the etching process is limited to one direction (Fig. 10.2b). This is caused by the fact that during plasma etching, the ions and chemical radicals bombard the wafer surface perpendicularly, creating an accurate copy of the mask pattern in the underlying layer. Therefore, dry-etching techniques are currently used for most etching process steps. Extremely high-quality etching processes, with high selectivity, have been developed for most materials that are currently used in IC production. New process generations, however, require improved selectivity, uniformity, reproducibility and process control. Particularly

Fig. 10.2 The results of different etching methods

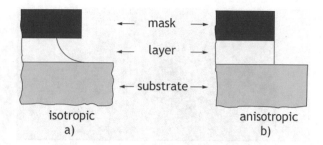

isotropic   anisotropic
a)   b)

the 3-D NAND flash memory technologies require high aspect ratio holes to be etched through a stack of more than 100 oxide and nitride layers! The focus on new etching techniques does not preclude further development of existing techniques.

10.2.3 Oxidation

Oxidation is the process of creating insulation areas or layers to electrically and physically isolate transistors and/or metal layers from each other.

A thermal oxide was used to isolate the transistor areas in conventional MOS ICs. In these areas, the oxide must be relatively thick to reduce interference of signals (polysilicon or metal tracks) that cross these areas. This thick oxide was created by exposing the silicon substrate to pure oxygen (dry oxidation) or water vapour (wet oxidation) at a high temperature of 900–1200 °C. The local oxidation of silicon (*LOCOS*) process was an *oxidation* technique, which had found universal acceptance in conventional MOS processes with gate lengths down to 0.5 μm. Silicon was substantially consumed at the wafer surface during this process. The resulting silicon-dioxide layer (LOCOS, Fig. 10.3a) extended about half below the original wafer surface and half above it. A disadvantage of the LOCOS process is the associated rounded thick oxide edge. This *bird's beak* is shown in the figure.

The formation of the bird's beak causes a loss of geometric control of the transistor channel width, which would become considerable as transistor sizes shrink. The use of LOCOS was therefore limited to 0.5 μm and larger CMOS technologies. An important alternative to this LOCOS isolation technique, already used in 0.35 μm CMOS technologies and below, is the *shallow-trench isolation* (*STI*). In this process trenches are etched in the isolation areas in the silicon wafer between the transistors. Then a dielectric material (usually silicon dioxide) is deposited to fill these trenches. So, STI is used to isolate transistors from each other but also borders the transistor channel. The STI process creates steep edges (Fig. 10.3b) and results in an accurate definition of the transistor channel width [32] but also creates stress in the transistor channel, which influences the threshold voltage.

Another important application of thermally grown oxide was the gate-oxide layer between the transistor gate and the substrate in conventional CMOS processes. A thinner gate oxide leads to a higher transistor current and higher circuit speed

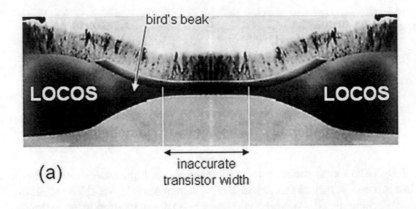

(a)   inaccurate
             transistor width

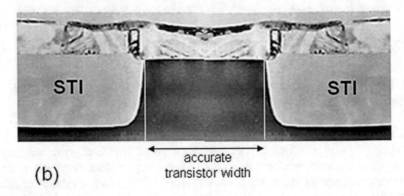

(b)   accurate
              transistor width

Fig. 10.3 Comparison of (**a**) conventional CMOS with LOCOS isolation areas and (**b**) state-of-the-art CMOS with STI isolation areas

(see Sect. 5.1). This *gate oxide* must be of extremely high quality, with a homogeneous thickness and very reliable. Defects such as *pinholes* (local areas in the gate oxide with oxide thickness below specification) and oxide charges have a negative effect on electrical performance and transistor lifetime. Because the gate oxide is only a few atoms thick (t_{ox} in Fig. 10.4), it is a continuous challenge for the industry to scale it further and/or find alternative ways to increase its capacitance. The use of SiO_2 for a gate dielectric layer thickness below about 2 nm causes *gate-oxide leakage* currents that may exceed a level of 1 A/cm^2. In the 90 nm node, SiO_2 was replaced by SiON as gate-oxide material, because of its higher permittivity for improved transistor performance, while keeping its leakage current within acceptable limits.

Further reduction of the gate-oxide thickness would severely degrade the performance as well as the reliability of the MOS transistor. As a result of this, the currently used polysilicon gate/SiON stack is replaced by a *metal gate* with so-called high-*k* dielectrics, which allows a large transistor drive current at a larger gate-dielectric thickness, causing much less gate leakage.

Fig. 10.4 Schematic cross
section of a MOS transistor

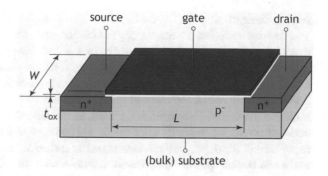

(bulk) substrate

Most advanced CMOS processes use *atomic-layer deposition* (*ALD*) to fabricate
the very thin gate-oxide layer. This has already shortly been described in Sect. 10. 2.1.

10.2.4   Diffusion and Ion Implantation

Diffusion and ion implantation are the two most commonly used methods to force
impurities or dopants into the silicon to create and optimise the nMOS and pMOS
transistors substrates, p-well and n-well, respectively, as well as the transistors
sources and drains and to adjust the transistors' channel conductivity, by adjusting
its threshold voltage.

Diffusion
Diffusion is often a two-step process. The first step is called pre-deposition and
comprises the deposition of a high concentration of the required impurity (dope).
The impurities penetrate some tenths of a micrometre into the silicon, generally at
temperatures between 700 °C and 900 °C. As the diffusion strongly depends on
temperature, each different diffusion process requires individual calibration for different processing conditions. During the diffusion process, silicon atoms in the lattice are then substituted by atoms from group three or five from the periodic system
of elements (*dopants*; see also Sect. 8.2).

The second step is called drive-in diffusion. This high-temperature (>1000 °C)
step forces the impurity deeper into the wafer, decreases the surface impurity concentration, creates a better homogeneous distribution of the impurities and activates
the dopants. This drive-in diffusion is not only in the vertical direction; it also causes
an identical lateral diffusion, which requires larger distances between individual
transistors. The continuous need for increased transistor density and doping profile
accuracy made diffusion techniques lose popularity during the 1990s. Ion
implantation has since then become the most popular method for introducing impurities into silicon.

Ion Implantation

The *ion implantation* process is quite different from the diffusion process. It takes place in an ion implanter, which comprises a vacuum chamber and an ion source that can supply phosphorus, arsenic or boron ions, for example. The silicon wafers are placed in the vacuum chamber, and the ions are accelerated towards the silicon under the influence of electric and magnetic fields. In fact, the implanter shoots the ions into the wafer. The penetration depth in the silicon depends on the ion energy and can be very accurately controlled. This makes ion implantation much more reproducible for doping silicon than classical diffusion techniques. It allows a large variety of doping profiles that are not possible with diffusion. With diffusion techniques, the maximum impurity concentration is almost always at the surface. The ion implantation technique, however, can be used to selectively create profiles with peaks below the wafer surface. The concentration of impurities decreases towards the wafer surface in these '*retrograde profiles*'.

The use of ion implantation in the formation of source/drain regions, particularly of the source and drain extensions becomes increasingly challenging as these junctions become very shallow (sub 10 nm shallow regions) in advanced CMOS processes [33]. The doping concentration does not increase with scaling. Only the energy during implantation must be adjusted to create those shallow junctions.

10.2.5   *Planarisation*

The growing number of processing steps, combined with a decrease in feature sizes, results in an increasingly uneven surface. For example, after completing the transistor fabrication steps, an isolation layer is deposited before the metal layers are deposited and patterned. The height variations of the underlying surface are replicated into this isolation layer. This unevenness introduces two potential problems in the fabrication process. When the first metal layer is directly deposited onto this layer, its thickness can dramatically vary with the variations in surface topology. Its thickness will be reduced at elevated steps, thereby increasing the metal resistance and reducing its reliability. Secondly, as already discussed in the lithography section, new lithography tools allow a smaller *depth of focus* (*DOF*), tolerating only very small height variations. During imaging, these variations can introduce focus problems at the elevated and low areas. Therefore, all CMOS processes use several *planarisation* steps, which flatten or 'planarise' the surface before the next processing step is performed.

CMOS technologies below 0.25 μm use *chemical mechanical polishing* (*CMP*). CMP is based on the combination of mechanical action and the simultaneous use of a chemical liquid (slurry). It actually polishes the surface. Figure 10.5 shows a simplified overview of the CMP process. The *slurry* contains polishing particles and an etching substance. A polishing pad together with the slurry planarises the wafer surface. Because CMP is also based on a mechanical action, it offers a better overall planar surface. Measures to prevent planarisation problems during fabrication of the

Fig. 10.5 Schematic
overview of the CMP
polishing process

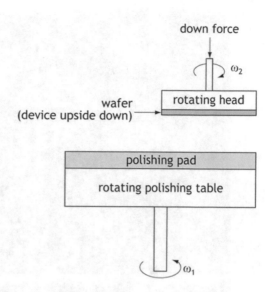

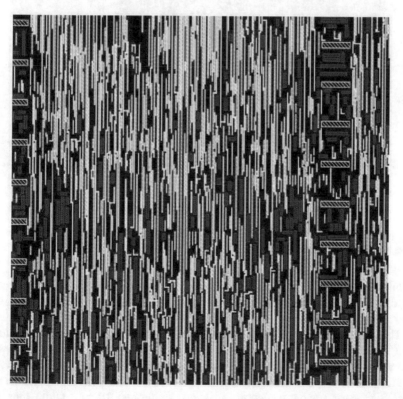

Fig. 10.6 Example of a more homogenous metal pattern distribution by using tiles (purple)

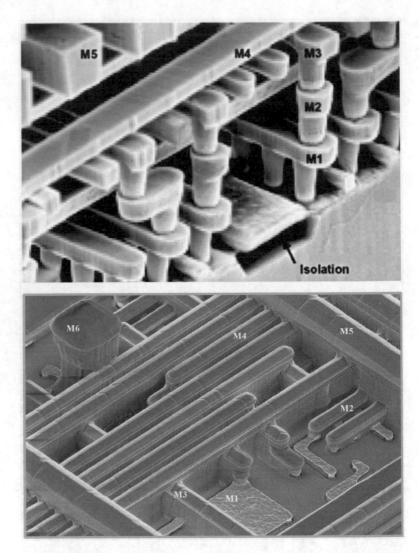

Fig. 10.7 Cross sections of CMOS back end, showing the potentials of CMP planarisation. (Source: NXP Semiconductors)

metal interconnect layers include the creation of *dummy metal patterns* in scarcely filled metal areas. The idea is to provide masks with an almost uniform density of their metal patterns to create a more homogeneously planarised surface.

These dummy metal patterns, also called *tiles*, are automatically included during chip finishing. Figure 10.6 shows an example of the use of *tiling* to achieve an improved metal distribution for optimised planarisation. Figure 10.7 shows the result of the CMP planarisation technique in a multi-metal layer process. Advanced CMOS processes show topology variations of around 10 nm, while they may increase to 80 nm close to the wafer edges. More on CMP can be found in [34].

10.3 Basic MOS Technologies

The previous sections illustrated that MOS processes mainly consist of several basic steps that are repeated many times. In advanced CMOS processes, the total number of all process steps is close to 1000.

The first MOS ICs were built from nMOS transistors only. We will therefore start with a short discussion on a basic nMOS process with just five masks. Understanding the basics of this simple process enables a smooth transition to the basic CMOS process. With the exception of some new steps, these CMOS processes are just an extension of the basic nMOS process presented here. Finally, a nanometre CMOS process is presented, and the associated fundamentally new steps are discussed. The section is concluded with a quantitative discussion of CMOS technology options beyond 16 nm.

10.3.1 The Basic Silicon-Gate nMOS Process

An nMOS process which uses a mere five masks is explained with the aid of Fig. 10.8. The right side in the figure shows several layout patterns corresponding to the successive process stages as illustrated by the process cross sections, which are through the dotted centrelines in the respective layouts. It illustrates the creation of the layout and fabrication of only one single transistor and its connections. It should be clear that today's ICs may contain several billion transistors, all fabricated simultaneously by the same process steps.

First, an oxide layer is grown on the bare silicon wafer, as a result of exposing the wafer to an oxygen environment. Next, the oxidised silicon wafer is coated with a silicon nitride (Si_3N_4) layer, as shown in Fig. 10.8a.

The first mask, which we call the ACTIVE mask, is used to define the transistor areas. Its colour is green. Only one green rectangle is drawn, meaning that we will only show the creation of a single transistor, but an ACTIVE mask may contain millions to billions of these rectangles, depending on the complexity of the chip. Everywhere outside the green areas, there will not be any transistor created, and thus these areas will become isolation areas. So, the ACTIVE mask is used to define the pattern in the nitride layer corresponding to substrate regions where transistors should be formed (Fig. 10.8b). In the conventional nMOS process, the wafer is then oxidised to produce the LOCOS isolation areas as shown in Fig. 10.8c. The resulting thick oxide only exists at places that were not covered by the nitride. After the removal of the remaining nitride and oxide layer of Fig. 10.8b, a thin oxide layer is grown in these areas (also in Fig. 10.8c). This *gate oxide* largely determines the properties of a MOS transistor. Gate oxidation is therefore one of the most critical processing steps. Its thickness is between 1 and 7 nm, depending on the process generation and the target chip application.

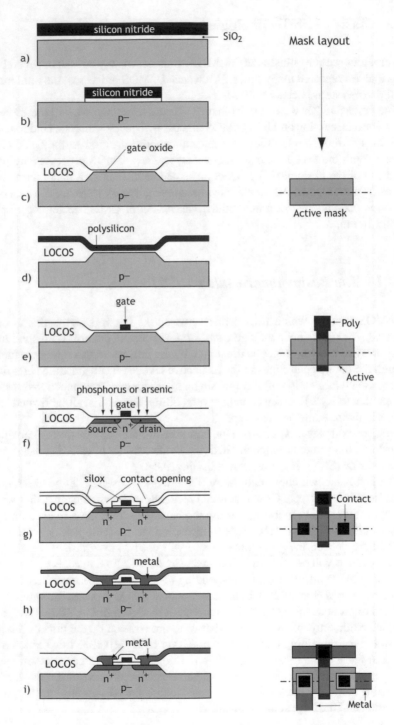

Fig. 10.8 The basic silicon-gate nMOS process with LOCOS Isolation (Detailed descriptions of the individual sub figures (**a** to **i**) can be found in the text)

Next, a polysilicon layer of about 100–400 nm thickness is deposited (Fig. 10.8d). Subsequent phosphorus diffusion is then used to dope the polysilicon for better conduction. Then the POLY mask, which is the second mask, is used to define the polysilicon pattern via masking, photolithographic and etching steps. The situation now corresponds to Fig. 10.8e. The polysilicon is used both as MOS transistor gate material, where it lies on the thin gate oxide, and as an interconnection layer, where it lies outside the active areas on thick oxide (LOCOS).

With a phosphorus (P) or arsenic (As) implant, the n^+ sources and drains are then created (Fig. 10.8f). They are actually 'shot' (implanted) into the wafer, with the LOCOS and polysilicon acting as a barrier. The edges of the n^+ areas are defined by the LOCOS on one side and the polysilicon gate on the other. Source and drain areas are thus not defined by a mask but are self-aligned with respect to the location of the gate. Below the polysilicon there is no n^+ dope. Instead, this region, in the top of the substrate below the gate and between the n^+ source and drain, is the transistor channel. The width of the green ACTIVE region below the gate, in the layout of Fig. 10.8e, is the *channel width*. The width of the polysilicon is the *channel length*.

Then, a new oxide layer is deposited on the wafer. This SILOX layer is about 200–800 nm thick, again depending on the technology node. Figure 10.8g shows the situation after contact holes have been etched in this layer according to a pattern in the CONTACT mask. This is the third mask used in this process. In these etched openings in the SILOX layer, connections can be made to the sources, drains and gates of the transistors. Next, a metal layer is deposited (Fig. 10.8h) all over the wafer. The METAL mask, which is the fourth mask in this simple process, is used to define the pattern in the metal interconnect layer. The used metal can be aluminium, tungsten or copper. The metal always connects to a transistor source, drain or gate through a contact hole. Basically, the processing is now complete as shown in Fig. 10.8i, which also shows a layout of the transistor and its connections. A *layout* is a kind of top view of the physical circuits on the chip and represented by the different mask layers.

As a final step, the entire wafer is deposited with a plasma-nitride *passivation layer*. This *scratch-protection layer* protects the integrated circuit from external influences, such as scratches, humidity and chemicals. Figure 10.8i shows the situation before deposition of the scratch protection. With a final mask, called NITRIDE, the scratch protection is etched away at the bond pad positions to be able to make wiring connections from the chip to the package. This NITRIDE mask and the associated processing steps are not drawn in the figure, because the bond pads are usually only located at the periphery of the chip.

In summary, the mask sequence for the considered basic silicon-gate nMOS process is as follows:

- ACTIVE   Definition of transistor areas
- POLY   Definition of the pattern in the polysilicon layer, which serves as transistor gate in the transistor areas and as local interconnection elsewhere
- CONTACT   Definition of contact holes between metal interconnections and the transistor sources, drains or gates

- **METAL**   Definition of the interconnection pattern in the metal layer
- **NITRIDE** Definition of openings in the scratch-protection layer, to enable bond pad connections through bond wires with package leads.

This basic simple and conventional nMOS process, used during the 1970s, only contained five masks. Logic gates built from only nMOS transistors showed relatively large power consumption when their output represents a logic '0'. This is because the nMOS transistor that is connected to the supply rail cannot fully be switched off, in this state, causing a permanent current from supply to ground while keeping the '0' output state. This was not so much of a problem, because nMOS ICs at that time did not contain many transistors. Around the early 1980s, the number of transistors on such chips became so high that the power consumption of nMOS chips reached the maximum power consumption of the cheap plastic packages, which was in the order of one to several watts. This was the main reason that the industry moved during the early 1980s to *complementary MOS (CMOS)* technology that enabled the use of both n-type and p-type transistors. The nMOS transistor that was connected to the supply rail in nMOS logic gates is in CMOS logic gates replaced by a pMOS, which can be fully switched off, when the output is in the '0' state. This means that a CMOS logic gate hardly consumes power when it does not change its state. The move from nMOS to CMOS reduced the power consumption roughly by a factor of 15.

The following sections show the additional steps to create a basic and an advanced CMOS process, successively.

10.3.2   The Basic Complementary MOS (CMOS) Process

CMOS circuits and technologies are more complex than their nMOS counterparts. In addition, a CMOS circuit contains more transistors and would therefore be larger than its nMOS equivalent. Both n-type and p-type transistors are integrated in CMOS processes. Figure 10.9 illustrates a cross section in a simple CMOS process.

Section 8.3 showed that nMOS transistors are fabricated on a p-type *substrate* and pMOS transistors on an n-type substrate. When using the same p⁻substrate as with the nMOS process in Fig. 10.8, we can still built the nMOS directly into this substrate (left side of Fig. 10.9). However, we need to add a so-called *n-well* or *n-tub* that serves as a substrate for the pMOS transistor (right side of Fig. 10.9). This is at the costs of an additional n-well mask and a few processing steps.

Because nMOS transistors have n-type sources and drains and pMOS transistors require the complementary p-type sources and drains, separate masks and process steps are used for their implantations: NPLUS mask for n⁺-implant in nMOS transistors in the substrate and PPLUS mask for p⁺-implant in pMOS transistors in an n-well. The *back end* of the process: the formation of the contacts and metal layers is similarly done as in the previously discussed simple nMOS process. So, in fact, a

Fig. 10.9 The basic CMOS process with LOCOS isolation

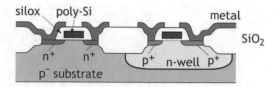

basic CMOS process requires only three more masks than a basic nMOS process: n-well, NPLUS and PPLUS.

Modern CMOS processes use 25 to more than 60 masks. Basically, these processes are all extensions of the simple CMOS process described above. VLSI and memory fabrication processes now use channel (gate) lengths (= distance between source and drain) of 10–350 nm and offer several levels of polysilicon and/or metal. These multiple interconnection layers facilitate higher circuit densities. The next section discusses a state-of-the-art nanometre CMOS process.

10.3.3   An Advanced Nanometre CMOS Process

Compared to the basic CMOS process discussed before, more advanced nanometre CMOS processes, with sub-100 nm channel lengths, include several major different processing steps. Figure 10.10 shows a cross section of such a process.

As a result of the continuous scaling of the transistor sizes, their current capability has been constantly under pressure. Therefore, the transistors themselves require several additional processing steps to enable further scaling of their physical sizes. Particularly the formation of the drain extensions and the halos around the transistor sources and drains are needed to suppress so-called short-channel effects that threaten a reliable transistor operation. Currently the wells are implanted instead of diffused and are called *retrograde wells*. All these additional steps are required to optimise transistor operation to maintain or improve their performance and reliability. A discussion of these steps is beyond the scope of this book but can be found in [2].

Next to these steps, current technologies must also support an increasing variety of application requirements. Therefore CMOS processes may contain different versions of both nMOS and pMOS transistors. MOS transistors with a low threshold voltage (*LV$_t$-transistors*) have a higher current capability and are used in high-speed circuits but suffer from relatively large leakage currents. On the opposite, MOS transistors with a high threshold voltage (*HV$_t$-transistors*) are used in medium- and low-speed circuits but have one to two orders of magnitude less leakage currents. For this reason most advanced CMOS processes are so-called dual-V$_t$ processes and support both high-speed and low-leakage circuits on the same chip. While most ICs run at voltages close to 1 V, their interface circuits (I/Os) usually operate at higher voltages, e.g. 1.8 V or 2.5 V, and require transistors with longer-channel length, larger oxide thicknesses and higher threshold voltages than the one used within the digital and memory cores. Analog circuits may use these transistors as well.

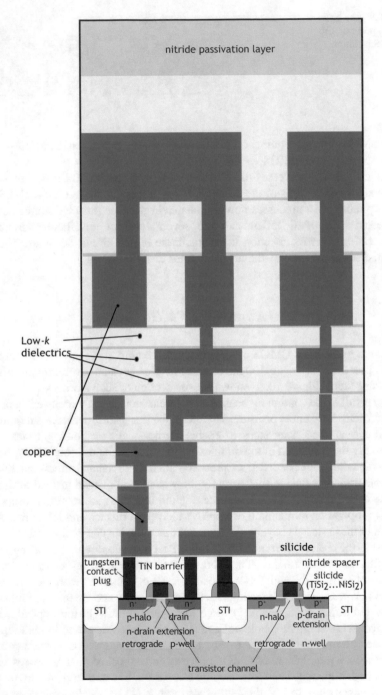

Fig. 10.10 An advanced nanometre process with STI isolation

In the back end of the process, the aluminium interconnect layers and contacts have been replaced by copper, while the number of these interconnect and via layers has increased dramatically. A via *layer* is a layer in between two metal layers, with vias on those positions in the layer where contacts between successive metal layers are required.

For advanced CMOS processes, these additional process steps and additional devices in combination with the large number of via and metal interconnect layers caused an increase in the number of mask layers to between 30 and 40, compared with only 5 mask layers for the conventional basic nMOS process, with which we started this chapter.

10.4 CMOS Technology Beyond 32 nm

Approaching the end of *Moore's law*, by reaching the physical limits of scaling planar CMOS devices, has challenged both process and design engineers to create solutions to extend CMOS technology scaling towards sub-10 nm feature sizes. An efficient integrated circuit is the combination of optimum operation of both the devices (transistors) and their interconnections. Local circuit speed is dominated by the devices (transistors' driving currents), while the global speed is dominated by a combination of the devices and interconnects (signal propagation). There are several issues related to the future scaling of the devices and interconnects.

10.4.1 Devices (Transistors)

The transistor's driving current depends heavily on its threshold voltage and carrier mobility. Scaling introduces several mechanisms that reduce this mobility, directly or indirectly. The conventional way of increasing the transistor current is to reduce the gate-oxide thickness. But with oxide thickness values (far) below 2 nm the transistor exhibits relatively large *gate-leakage currents*, which increase with a factor of ten for every 0.2 nm further reduction of the oxide thickness. Low leakage is a must today, because it greatly determines the *standby time* of most mobile electronic gadgets. A *high-k gate dielectric* (hafnium oxide, zirconium oxide and others) is therefore a must to continue device scaling with an affordable leakage budget. It was very complex to find the right combination of high-k gate dielectrics with the right gate electrode to offer optimum transistor performance and reliability. Intel developed a so-called gate-last CMOS process, in which the sources and drains are created before the gate electrode and introduced the Penryn dual-core processor with 410 million transistors in 45 nm CMOS with high-k gate dielectrics and metal gate already in 2008 [31]. Many other semiconductor vendors have recently installed similar technologies in their production lines.

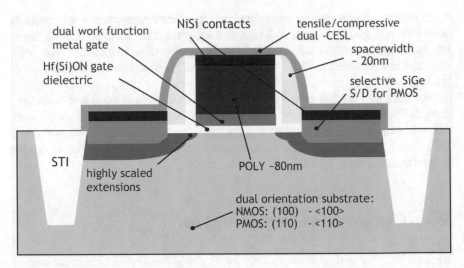

Fig. 10.11 Potential technology options for performance boost of MOS devices. (Source: NXP Semiconductors)

Another way of increasing the transistor current is to improve the *channel mobility* by building stress/strain in or close to the transistor channel. In unstrained nanometre CMOS processes, the average mobility in pMOS transistor channels is more than two times lower than that in nMOS channels. Therefore, the improvement of the pMOS transistor mobility was given more priority by building stress into their architecture. Figure 10.11 shows several process options to enhance the carrier mobility for improved nMOS transistor performance. This is also referred to as *strained silicon*. To achieve the best lateral mobility improvements, the strain should be compressive for the pMOS transistors and tensile for the nMOS transistors [35].

The carrier mobility in the channel is also related to their physical crystal orientation which was discussed in Sect. 8.4.4. The optimum combination of stress and device orientations has driven and will still further drive the transistor currents to higher values than available in today's high-volume CMOS processes.

Other alternatives to increase the transistor current capability include a double-gate or FinFET transistor. In a *double-gate transistor* (Fig. 10.12a), the transistor body is still lateral but embedded in between two gates, a bottom gate and a top gate which means that there are two parallel channels contributing to the total current of the device, which now behave as two parallel transistors.

In the example *FinFET* architecture (Fig. 10.12b), a narrow vertical substrate (fin), about 7–10 nm thick, is located on top of a *buried-oxide layer* (*BOX layer*) and then covered with a thin high-*k* dielectric layer. Then a thin metal layer with a poly silicon cap is formed, covering the gate-oxide areas surrounding the fin at all sides: left, top and right side. This device will operate with a higher drive current, due to the parallel current paths. Most FinFET processes, today, are based on bulk-silicon wafers, which do not include the BOX layer.

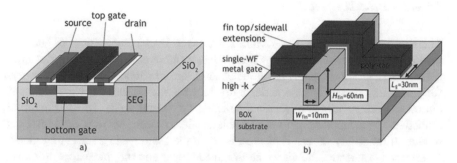

Fig. 10.12 (**a**) Double-gate transistor and (**b**) FinFET. (Source: NXP Semiconductors)

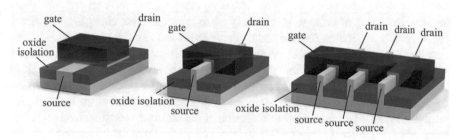

Fig. 10.13 Traditional planar CMOS transistor (left) and FinFET (middle), both on a bulk silicon wafer, and three FinFETs in parallel controlled by the same gate (right)

Double-gate and FinFET devices are also called multi-gate FET or *MuGFET*. Both help to control leakage currents and reduce short-channel effects. As they do not exhibit doping fluctuations, their operating properties are much better.

Many companies have adopted FinFETs in their (sub-) 16 nm process nodes. Most FinFETs today, however, are built on bulk silicon wafers for improved compatibility with the planar CMOS process and to reduce cost. In an example of a bulk CMOS 22 nm FinFET process [36], the formation of the fins is similar to the formation of the active areas in a planar CMOS process (Fig. 10.13, left), by etching trenches (STI) in the silicon wafer and filling them with isolation oxide, after which the wafer is planarised. In a FinFET process, this STI oxide is then etched back (recessed) so that the fins reveal (Fig. 10.13, middle). Further process steps to create the gate stack and interconnections are similar to those in a high-ε/metal gate process and will not further be discussed. Most circuits require more current than just that of one FinFET. In these cases, sources are connected to each other, and drains are connected to each other (Fig. 10.13, right) to create several FinFETs in parallel, acting as just one transistor.

10.4.2   *Interconnects*

There are several reasons why CMOS ICs need an increasing number of interconnect layers. Every new technology node offers us more transistors with a two times
higher density. This requires more metal resources to support the increasing need
for connecting these transistors. Secondly, they require a more dense power distribution network to be able to supply the increasing current needs. This limits further
reduction of the metal height because this would increase the track resistance and
reduce its reliability. There is also an issue in the scaling of contacts and vias (a
contact is a connection between the first metal layer and the transistors; a via is a
connection between two successive metal layers). Since their number and *aspect
ratio* (height/width ratio) increase with scaling, while their sizes decrease, they are
becoming a very important part in the determination of the global chip performance,
reliability and yield. Because of the increasing currents, the contacts and vias show
an increasing loss of voltage, particularly when the signal line changes many times
from metal layer. It then forms a serial chain of vias and metal tracks in these layers.
Finally, due to the high aspect ratios, there is an increased chance for bad contacts
or opens, which will affect the yield. Already today, design for manufacturability
(DfM) has become an integral part of the design flow to support yield-improving
measures (see also Chap. 12). Most of the further improvements of the interconnect
network has to come from further reduction of the dielectric constant (*low-k dielectrics*) of the *inter-level dielectric* (*ILD*) layers between the metal layers and between
the metal lines within one layer (Fig. 10.10). During the last two decades, this
dielectric constant has gradually reduced from 4 to 2.5. To achieve a low-k dielectric, Intel uses so-called airgaps between metal lines in their Broadwell processor in
2014. It is expected that dielectric constant will further reduce to close to 2, but it
still needs many innovations to guarantee sufficient reliability. For technology nodes
of 10 nm and below, Intel is expected to use cobalt in the bottom two interconnect
layers. It is claimed to yield a twofold reduction in via resistance a five- to tenfold
improvement in electromigration [37].

10.5   Conclusion

It is clear that the realisation of optimum electronic systems is based on a perfect
match between the substrate (wafer), the transistors and their interconnections. The
increasing number of application areas has led to a large variety of technology
options to support the different requirements of high-speed, high-density and low-
power products. This has increased the number of mask layers from five, during the
1970s, to more than sixty today.

It should also be clear from this chapter that the continuous scaling of integrated
circuits has brought us close to the physical limits. To achieve further improvements

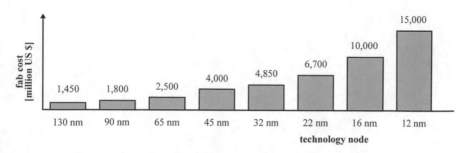

Fig. 10.14 Cost of a 300 mm wafer fab as a function of the feature size

in density and speed and reduction in power requires a continuous change in transistor and interconnect architectures.

This requires huge investments in the manufacturing chain: mask making, lithography, production process, testing and packaging. A very interesting website is that of the International Technology Roadmap for Semiconductors (*ITRS*) [38], where a lot of data on the expected roadmap for the next 10 years can be found. Unfortunately, the latest 'traditional' ITRS roadmap was published in 2013.

Because of the changing semiconductor markets, from computer and consumer to mobile, automotive and IoT, the semiconductor industrial association decided to change the focus of the ITRS towards seven topics and continued with an ITRS 2.0 version. With the end of Moore's Law at the horizon, ITRS 2016 has become the final roadmap and will be replaced by a new initiative, named the *International Roadmap for Devices and Systems (IRDS)*, which now also include *beyond CMOS* or *more than Moore* devices. This is also the result of the fact that there are only a few semiconductors companies left that take the lead in the creation of new CMOS process nodes.

Currently, a new fully equipped 12 inch (300 mm) fab could take output to several hundred thousand to a million wafers per month. To ramp up a fab to volume production in a 22 nm requires a time frame of about two and a half years and a budget of more than $10 billion. This has prompted many semiconductor companies to outsource most of their production and become 'fab-lite' or maybe even totally fab-less. This trend is certainly continued beyond the sub-20 nm technology nodes.

Finally, the evolution of wafer fab cost is shown in Fig. 10.14.

Chapter 11
Chip Performance and Power

11.1 Introduction

Although already used in the 1970s, it took until the mid-1980s before CMOS became the leading-edge technology for VLSI circuits. Prior to that time, only those applications that really required the low-power features of CMOS were designed in it. Most examples, then, were battery-supplied applications, such as wristwatches, pocket calculators, portable medical devices (hearing aids and implantable heart controls) and remote controls.

Power reduction is a must and a real differentiator in almost every application today. Therefore a short overview on the necessity of power reduction will be discussed first.

Basically, a VLSI chip is just a bunch of transistors, and the way they are interconnected determines its functionality. The continuous scaling of the feature sizes on a chip according to Moore's law not only resulted in extremely high transistor densities but also improved the performance and reduced the power consumption of electronic systems. This chapter starts with a discussion on the impact of scaling on the basic elements: the transistor and the interconnections. Next the consequences for the overall performance and power of nanometre CMOS ICs are discussed.

The fast development of a large variety of mobile devices not only forced a 'less-power' design attitude but also increased the drive for significant battery performance improvements. A review of battery basics will therefore conclude this chapter.

11.2 The Need for Less Power Everywhere

Traditionally, power reduction was only a requirement for battery-operated products. In fact, the phrase 'low-power design' was directly referring to the design of chips for mobile markets. It suggests that other applications did not need a

© Springer International Publishing AG, part of Springer Nature 2019   189
H. Veendrick, *Bits on Chips*, https://doi.org/10.1007/978-3-319-76096-4_11

Fig. 11.1 A universal
CPU cooler fan with
heatsink. (Source:
StarTech)

less-power design attitude. Therefore we should not talk about low-power design but use the more generic phrase 'less-power design' which refers to the necessity of power reduction at every level of design and process technology in every chip. A few examples will be given below.

Around the early 1980s, the average *power consumption* of consumer ICs exceeded the maximum allowed power figure of 1 or 2 watts of cheap plastic packages, such as DILs and DIPs (Fig. 14.1)). This was one of the main reasons for moving from nMOS to CMOS technology at that time. This move reduced the average chip power consumption with about a factor of 15. Due to increasing number of transistors and switching frequencies, the average CMOS chip exceeded that power level again, already in the early 1990s. However, since then, there is no alternative technology. The maximum power figure of the cheap plastic packages remains one of the main driving forces for less-power CMOS design in consumer applications.

From the 1970s until today, the number of transistors on high-speed microprocessors (also called *central processing units*, *CPUs*) such as the ones from Intel and AMD increased from only a few thousands to more than five billion. In the same period, the maximum on-chip clock frequencies increased from several megahertz (MHz) to several gigahertz (GHz). As a consequence, the power consumption of these ICs increased from less than 1 W to more than 100 W, despite the move from nMOS to CMOS in the early 1980s. Figure 11.1 shows a cooler fan with heatsink for high-speed microprocessors to 'drain' the heat that they produce.

It is clear that more power will lead to a higher fan volume, which would certainly become a huge problem in (mini) laptops and tablet computers. Power reduction in these high-performance microprocessor chips is therefore even a stronger requirement than in battery-operated products. An additional benefit, here, is that less power also means less noise of the fan.

The increasing number of portable applications is a third driving force for less-power CMOS. It requires access to powerful computation at any location. In the consumer market, we can find examples such as games, smart phones, tablets, MP3 players, digital cameras, e-readers, GPS systems and DVD players. In the PC market, an increasing percentage of computers is sold as (mini) laptop. Digital cellular telephone networks, which use complex speech and video compression algorithms, form a less-power CMOS application in the telecommunication field. The emerging multimedia market will also enable many new products in the near future. Currently, we see the portable full motion video and graphics as examples of such low-power applications.

Another important driving force for less power is the future system requirement. In a 10 nm CMOS technology, more than 1000 billion transistors can be packed on a 20 cm by 20 cm board with very high-density packaging techniques, such as MCM, system in a package (SiP) and system on a package (SoP). Current power levels are not acceptable for these systems.

In all electronic products, from devices to systems, less power also leads to simpler power distribution, less supply and ground bounce and a reduction of electro-migration and electromagnetic radiation levels.

Finally, integrated circuits will increasingly be used to help secure a *sustainable environment*. One of the biggest requirements to achieve this is to limit overall energy usage and cut waste. A *smart grid* enables communication between the supplier and the consumer to optimally match the overall power supply (electricity, gas) to the power demand. It uses *smart metering*, such that the consumers can decide how and when they use energy, e.g. to run appliances at times that demand and price are low.

All of the above show that a less-power design attitude should be common in the total chip design trajectory, from system level through layout level to process level, because it allows cheaper packages, it increases robustness and reliability of integrated circuits and systems and it lengthens both the active and standby times of mobile gadgets. Finally it helps to reduce global warming.

There are many options, both in design and in process technology, to reduce the power consumption of integrated circuits. A discussion of these is beyond the scope of this book but can be found in [2].

11.3   Impact of Scaling on Power and Performance

11.3.1   Transistor Scaling

Figure 11.2 shows the evolution of voltage scaling over the last couple of decades.

For many technology generations in the past (before the 1990s), the supply voltage has been constant and equal to 5 V. The scaling process over that period of time was called *constant-voltage scaling*, referring to only the scaling of transistor sizes and interconnections while keeping the chip supply voltage constant. This involved technologies above and including 0.8 μm (=800 nm).

This was followed by nearly a decade of *constant-field scaling*. In this period the transistor channel scaled from 0.8 μm (5 V) through 0.5 μm and 0.35 μm (both 3.3 V), 0.25 μm (2.5 V), 180 nm (1.8 V) to 120 nm (1.2 V). So, during this scaling period, the scaling of the voltage was equal to the scaling of the transistor channel length, meaning that the field in the channel remained constant. This constant-filed

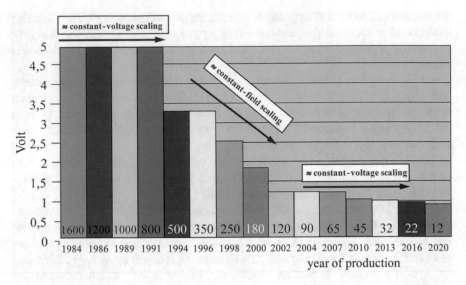

Fig. 11.2 The voltage scaling evolution for CMOS processes [38, 39]

scaling period was extremely beneficial for power reduction. For example, assume that a signal processor core in the 250 nm CMOS node consumed a certain power to perform a dedicated signal processing function. After scaling this core to the 180 nm node, it will consume about 2.5 times less power at half its size, when executing the same function at the same speed. Particularly this constant-filed scaling period has enabled many different electronic gadgets, such as GPS, MP3 player, mobile phone, digital camera, etc., to transfer from individual voluminous power-consuming electronic devices into one integrated trendy mobile device.

After the millennium, the voltage has hardly been reduced further and is still close to 1 V. Both the reduced operating margins of on-chip analog circuits and static random-access memories (SRAM) are a bottleneck to further reduction of the voltage. The scaling during the last decade could thus also be characterised as constant-voltage scaling (Fig. 11.2). Although the density increase remained about the same with further scaling, the power reduction and performance improvements were much less for the sub-90 nm nodes. This has forced the high-performance processor vendors to move away from the GHz speed race to a multicore processor race.

There are four important effects of scaling the transistor sizes and voltages:

Reducing Transistor Performance Improvement
A transistor current is just a flow of charge carriers (electrons or holes) through the transistor channel. The faster they can move, the higher their *mobility*, the more current a transistor can handle and the faster the circuit becomes. However, due to the continuously reducing transistor channel widths and lengths, it becomes increasingly difficult for the charge carriers to cross the channel. In other words, the mobility of these carriers reduces. This reduction can only be compensated by changing

the transistor architecture and manipulating the doping concentration in the channel of the transistor, as discussed in the previous chapter. This is at the cost of additional masks and processing steps.

Reducing Dynamic Power Consumption of Switching Devices and ICs

Power consumption of a chip is mainly caused by the switching of its transistors. This *switching power* is the largest contribution to the total power consumption of a chip. It is expressed as $P = CV^2f$, where C is the total load capacitances of all switching nodes, V is the supply voltage and f represents the switching frequency. When we scale the devices and interconnections, the load capacitance will reduce accordingly. When we scale the voltage, the power is reduced with its square, meaning that voltage scaling is one of the best measures to reduce power consumption. Finally, faster circuits allow higher frequencies leading to an increase in power consumption. This was one of the limitations to further increase the clock frequency of high-speed microprocessors.

Increasing Transistor Leakage

The continuous decrease of the transistor sizes in combination with the reduction of the supply voltages has made the transistor switching behaviour far from ideal. The voltage scaling from 5 V to roughly 1 V has also forced to reduce the transistor threshold voltage V_t in order to maintain sufficient transistor current (see the transistor current expression in Sect. 5.1). This threshold voltage has become so low that even when the transistor is 'switched off', it is leaking a small current. This *sub-threshold leakage current* increases by a factor close to 18, for every 100 mV reduction of the threshold voltage. If a chip would only contain a few thousand transistors, this *leakage current* would not be much of a problem. However, today's ICs may have hundreds of millions to several billion transistors, which together carry a relatively large leakage current, leading to a relatively large *standby power*. This is particularly a problem in handheld devices, where the battery-up time is very much dependent on the standby power consumption. Solutions to this problem exist in the form of additional fabrication steps in combination with design measures. This has been achieved by offering transistors with different thresholds voltages (dual-V_t processes) and leakage currents: low thresholds for high-speed functions and high thresholds for low-standby power functions. Also the various cores on a chip may run at different voltages. Individual cores may include large *power switches* to enable them to be completely switched off during standby periods. An advanced system-on-a-chip (*SoC*) may therefore consist of different *voltage* and *frequency domains* which make the design additionally complex and time consuming.

Decreasing Reproducibility of Transistor Performance

In Chap. 5 it has been explained that the current through a transistor is dependant on the number of doping atoms in the transistor channel. In conventional CMOS technologies, the transistor sizes were so 'large' that the average number of dopants in the transistor channel was several thousand per transistor. In advanced CMOS processes, with sub-30 nm channel lengths, the number of dopants can be even less than 100. This means that the fluctuation in dopants causes an increase in the spread

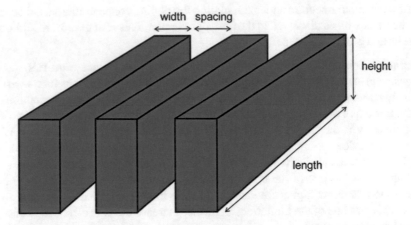

Fig. 11.3 Example of three metal wires on an advanced CMOS chip

in transistor currents. As a consequence, completely identical transistors may carry increasingly different currents. Particularly analog circuit performance is based on how well two equally sized transistors behave identical (match in behaviour). Because of this increasing *mismatch*, most analog circuits on an advanced CMOS chip do not use the minimum allowed transistor sizes but use larger channel widths and lengths to reduce the channel dopants fluctuations. They may also use a higher supply voltage.

These above four effects require additional process steps and/or design measures to compensate the disadvantages of further scaling of the transistors, which eventually lead to an increased design and fabrication complexity and cost.

11.3.2 Interconnection Scaling

The interconnections between the different components on a chip (transistors, analog blocks, digital blocks, memories, interfaces, etc.) are implemented as metal wires. These metal wires have a certain minimum width and a minimum spacing as defined by the technology node. Figure 11.3 shows three parallel metal wires, all with a minimum width and a minimum spacing to their neighbours. These wires could be part of a 16-bit bus, serving as an on-chip communication channel.

For educational purposes, the figure shows relatively short wires. However, due to the increased chip complexity, the length of these wires increases, while their height, width and spacing decrease. It takes some time for the signals to propagate over these wires from one point on a chip to another. This is called *propagation delay*. Even when the wire length is constant, the scaling of the width and spacing causes an increase of the signal propagation delay in the next technology node. This delay is proportional to both the resistance and the capacitance of the metal track.

The capacitance of a metal track is the ability of the track to store electrical charge. A larger capacitance means that it can store more electrical charge. To charge a capacitance to a certain voltage level requires an amount of energy, which is proportional to both the capacitance value and the voltage level. In other words, larger wire capacitances mean larger power consumption during switching.

The track resistance (R) increases with decreasing track width, while the track capacitance (C) to the neighbours increases with decreasing track spacing. As a result, the propagation delay $(\tau = RC)$ is increasing every new technology node. Figure 11.4 shows that the average signal propagation delay between two successive technology nodes increases by a factor of around 1.3. Was the delay over a 10-mm-long wire less than 0.5 ns in a conventional 0.5 μm CMOS technology, in 22 nm CMOS today, it would almost take 7 ns, which is more than ten times as much. This is quite opposite from what is demanded by the new systems: improved performance requires reduction of the propagation delay. The propagation delay can be reduced by two or three times by doubling or tripling the spacing between the metal tracks, respectively.

Another disadvantage of reducing the spacing between metal wires is the dramatic increase of interference from neighbouring signals. This so-called *crosstalk* introduces a lot of noise and reduces the so-called *signal integrity*.

Both effects, increasing propagation delay and reducing signal integrity, require additional design measures to guarantee robust chip operation at the required performance levels. Also in the process technology, new *low-k dielectrics* have been introduced to reduce the capacitance between the metal wires. The corresponding reduction of the dielectric constant k (ε_r in the figure) has already been incorporated in the diagram of Fig. 11.4.

Scaling of widths and spacings has caused the metal interconnections to dominate the IC's power, performance, reliability and signal integrity. Table 11.1 shows the increase in the average ratio between the metal wire capacitance and the transistor capacitance in digital circuits due to scaling. In advanced CMOS processes, the wire capacitance contribution is much larger than the transistor capacitance causing the wires to dominate both the chip performance and power consumption. Figure 10.10 (Chap. 10) shows an example cross section of a 45 nm CMOS circuit. The figure clearly demonstrates the increased dominance of the metal interconnect in nanometre CMOS processes. In the example of a nanometre CMOS SRAM, not only the increase of the mutual capacitance between two minimum-spaced wires is important but also the increase of the mutual capacitance between two minimum-spaced contacts or two minimum-spaced vias.

Figure 11.5 shows a 3-D cross section of an SRAM array. It requires an accurate 3-D extraction tool to enable proper simulation and prediction of the SRAM memory performance and power consumption.

The above has shown that both the scaling of the transistors and of the metal wires have a dramatic impact on the chip's performance and power consumption, which form an important drive for applying power reduction techniques at all levels of design: system, core, layout and even transistor design and interconnect usage.

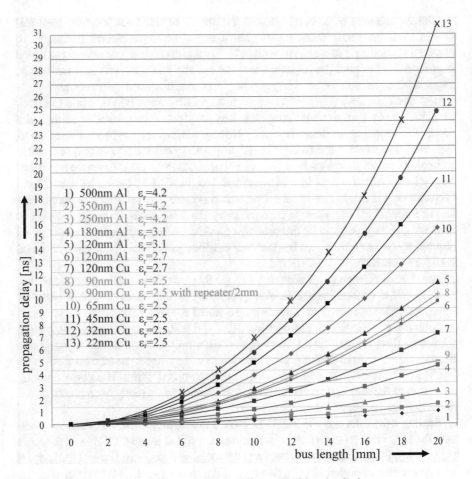

Fig. 11.4 Propagation delay of a metal track in different CMOS technologies

Table 11.1 Increasing
interconnect dominance

Technology	Ratio: wire load/fan-in
350 nm	30/70
250 nm	33/67
180 nm	36/64
130 nm	45/55
90 nm	54/46
65 nm	66/34
45 nm	75/25
32 nm	81/19
28 nm	83/17
22 nm	85/15
16 nm (FinFET)	78/22

The legend within the figure reads:

1) 500nm Al ε_r=4.2
2) 350nm Al ε_r=4.2
3) 250nm Al ε_r=4.2
4) 180nm Al ε_r=3.1
5) 120nm Al ε_r=3.1
6) 120nm Al ε_r=2.7
7) 120nm Cu ε_r=2.7
8) 90nm Cu ε_r=2.5
9) 90nm Cu ε_r=2.5 with repeater/2mm
10) 65nm Cu ε_r=2.5
11) 45nm Cu ε_r=2.5
12) 32nm Cu ε_r=2.5
13) 22nm Cu ε_r=2.5

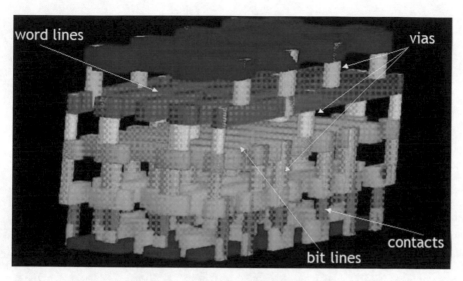

Fig. 11.5   3-D cross section of a nanometre CMOS SRAM array

11.3.3   *Scaling Impact on Chip Performance and Power Parameters*

Assuming that we continue to use bulk-CMOS wafers and that the transistor architecture does not change dramatically, then the trends in various performance parameters over the last couple of decades and their expected improvements are depicted in Fig. 11.6. The figure also assumes that the first year for volume production for the 32 nm and 22 nm nodes was delayed with respect to the 2 year cycle with which technology nodes were introduced before. The results are based on the evolution of the voltage scaling for low-standby power (LSTP) CMOS processes as shown in Fig. 11.2. The *scaling factor S* is defined as the average ratio between the minimum dimensions of transistors and interconnections in two successive CMOS technology nodes. Over the last four decades, this factor was close to $S = 0.7$.

The figure shows that the *constant-field scaling* era has been particularly beneficial for *power-efficiency* improvement, because it is inversely proportional with S^3, with S being the average *scaling factor* between two successive technologies. This is due to the combined scaling of the sizes and the supply voltage during that period. A new generation of electronic devices exhibited about two-and-a-half times more functionality for the same power needs, compared to its previous generation. In that same period of time, the subthreshold leakage power has increased by more than three orders of magnitude, which was a major drive to limit further scaling of the supply and threshold voltages. The diagram also clearly shows that below 100 nm,

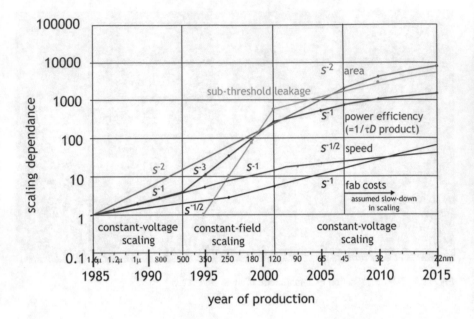

Fig. 11.6 Experienced and expected trends in various performance parameters in relation with the fabrication cost [2]

the improvements both in power efficiency and speed are only limited and that it will remain so in the future, while the fab cost are still expected to increase at least with the same trend.

The above-discussed trends have severe impact on the design of integrated circuits. For high-performance microprocessors, this has led to moving away from higher-frequency architectures towards multicore architectures. Chapter 1 in Part I has presented a couple of photographs of multicore processor architectures for several application domains. For vendors of most consumer and telecommunication ICs, not only the performance and functionality of an IC are differentiators but certainly also the power consumption. *Power management* at all hierarchy levels of design has therefore become a necessity for a successful introduction of a product into the market. The higher the level of design, the more impact it has on power saving [40]. This is also represented by the diagram in Fig. 11.7.

Standby and operation times of mobile devices are not only determined by the amount of power used by these devices but also by the battery size and capacity. The following section presents a summary on battery technologies.

Fig. 11.7 Influence of a
power reduction measure
at different design
hierarchy levels

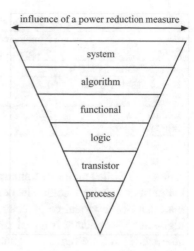

influence of a power reduction measure

system

algorithm

functional

logic

transistor

process

11.4   Battery Technology Summary

A *battery* is usually built from more than one cell, which can chemically store energy for a certain period of time. Based on the difference in the chemical process, we can distinguish two different types of batteries:

- Non-*rechargeable batteries* use so-called primary cells with a nonreversible chemical reaction and must be hand in as small chemical waste, when empty. These primary battery cells perform well in terms of charge capacity, charge storage and charge leakage but are less cost-efficient in mobile high-performance systems or systems that are always on.
- *Rechargeable batteries* use secondary battery cells, which deliver energy by transforming one chemical substance into another. This transformation is reversible in that it can be put back into its original chemical state during recharging. In battery-operated systems that need frequent replacement of the batteries, rechargeable batteries would be a more economically viable solution. But, in applications which need a long battery lifetime, e.g. a year for an electronic clock or remote control, most rechargeable batteries must be recharged at least every 3 months, due to their increased self-discharge, while a non-rechargeable battery may 'tick' for more than a year.

The growing diversity of battery-operated systems, combined with the increasing performance and longer battery lifetimes, requires improved battery energy

Table 11.2 Characteristics of rechargeable batteries

Battery type	Nominal cell voltage [V]	Energy/volume [Wh/L]	Energy/weight [Wh/kg]	Self-discharge rate [%/month]
Nickel-Cadmium (NiCd)	1.2	200	100	10–20
Nickel-Metal Hydride (NiMH)	1.2	400	200	10–30
Lithium-ion/ lithium-polymer	3.7	750	265	3–10

efficiency, while smaller weight and shrinking dimensions require a reduced number of stacked battery cells. The performance of cells in series is substantially worse than that of individual cells. A single-cell battery with both high cell voltage and high-energy efficiency is a real need in many applications. Advances in rechargeable battery technologies are aimed at improving the battery capacity per unit of volume while reducing their leakage currents.

Nickel-cadmium batteries have dominated the battery market for a long period of time, but they suffer from low cell voltage and low-energy efficiency (see Table 11.2). The nickel-metal-hydride (NiMH) batteries have replaced the nickel-cadmium (NiCd) ones in many applications, because of the higher-energy capability. Both the NiCd and NiMH types of batteries suffer from the so-called memory effect. In many applications, these batteries are recharged before they are sufficiently or completely empty. After many of such recharge operations, the battery starts suffering from a so-called voltage depression, also known as the *battery memory effect*, which reversibly degrades its energy storage capacity. They therefore need a periodic deep discharge to prevent this memory effect.

During the last decade, single-cell lithium-ion (Li-ion) and lithium-polymer (Li-pol) batteries have emerged as the more favoured choice. In a Li-pol battery, the lithium electrolyte is a solid polymer medium as compared to the organic solution in a Li-ion battery. They both offer a higher cell voltage and a higher-energy density (up to 600 Wh per litre). Because lithium is one of the lightest elements of the periodic system (it is element number 3), it helps to save weight, particularly in tiny handhelds. Moreover, the self-discharge rate is only 5% per month and they hardly exhibit the memory effect. A major disadvantage of the lithium batteries is their sensitivity to over (dis)charge, or short circuit, because this can cause them to ignite or even explode. Li-ion and Li-pol battery packs may therefore contain internal protection circuits to monitor its voltage to prevent battery damage and its temperature to disconnect from the application, in case it gets too hot. These batteries should therefore not be used in applications in which they could be exposed to high temperatures.

The self-discharge rates in Table 11.2 are related to traditional batteries that exist already for a couple of years. Over the recent years, low self-discharge nickel-metal hydride (LSD-NiMH) batteries have entered the market. These batteries should still retain 75–85% of their capacity after 1 year when stored at room temperature.

Although their capacity is less than that of the traditional ones, already after a few weeks of storage, their retained capacity exceeds that of the higher capacity traditional ones.

As electronics becomes more mobile, the demand for improved battery technology will continue to increase. Most of these applications are in the range of 10 mW (jpeg encoding in a cell phone) to 10 W (peak power in a mobile device). However, the incremental improvements in battery technology do not keep pace with this increase in battery demand, and, as such, it puts an additional burden on the shoulders of the IC design community by requiring a more intensive use of less-power design methods.

More information on battery technologies can be found in [41, 42].

11.5 Conclusion

The scaling of the transistor and interconnect sizes severely impact chip and system performance and power consumption. The power consumption of a chip can be reduced in various ways, but the largest power savings can be achieved by reducing the supply voltage. In this respect, the scaling process from 0.8 µm CMOS technologies to the 28 nm node, in which the supply voltage gradually reduced from 5 V to roughly 1 V, respectively, has had a huge impact on the total power reduction. This was the basis for the integration of a large variety of functional features (camera, MP-3, phone, games, internet access, blue tooth communication, audio, video, GSM, compass, etc.) into a single mobile gadget: the smart phone as well as the tablet computer.

The drive for further scaling had already moved away from speed increase towards density improvement. A scaling factor of 0.7 for the transistor sizes still enables the implementation of about twice the hardware on the same chip area. Already more than a decade ago, the high-performance microprocessor race in terms of gigahertz has made place for a race towards more processor cores on one chip.

The limited improvements in battery technology, combined with an increasing number of features and functions on a chip, have made power consumption a key differentiator in choosing the best chip solution to create the optimum system for mobile devices. Reduced power consumption is of equal importance for all other chip applications, since it allows small and cheap plastic packages thereby reducing the overall system sizes and cost.

The above has made clear that state-of-the-art chip design requires a 'less-power' attitude of all electronic designers at all levels of design: application level, system level, chip level and transistor level.

Much more detailed information about power sources and measures to reduce chip power consumption can be found in [2].

Chapter 12
Testing and Yield

12.1 Introduction

An integrated circuit can fall victim to a large variety of failure mechanisms. Ideally, the related problems are detected early in the manufacturing process. However, some only show up during the final tests, or even worse, they might not be identified before the chip is soldered on a customer's board.

So, when the first wafers from the first production batch arrive from the fab, it is a very exciting moment for the designer to put this *first silicon* on the tester. More than 60% of all first silicon is first-time-wrong, mostly caused by logical errors and timing failures. Finding the root cause of the failure may be very time-consuming. The cost of a *respin*, which includes a new mask set, the time needed to find and diagnose the cause of failure and a 3 month delay in market introduction, may easily exceed $3 million.

Testing, yield and failure analysis therefore have a substantial influence on the ultimate costs and quality of a chip. An overview of these topics is therefore a necessary part of this book.

Not every chip on a wafer will be fully functional. Failing chips may be caused by random defects during the fabrication process, by critical mask dimensions, by design errors or by limited or too small design margins.

The engineering and evaluation of first silicon until it is considered to be 'error free' happens to be a tough job. Programmable processors, for example, may be used in an almost unlimited number of different applications. It is therefore very difficult to guarantee even 'fifth-time-right' silicon for these kinds of ICs.

Even when a failure is detected during the testing of first silicon, it might take a considerable time before the cause of failure is located and proven. This is because complex ICs may contain several hundred million up to several billion transistors and only a few hundred to a few thousand of external pins (I/Os). It is therefore very complex to diagnose and locate an internal failure via a limited number of I/Os. Moreover, because of the increased number of interconnect (metal) layers, which limit physical access to the circuits below, physical probing of signals has almost

© Springer International Publishing AG, part of Springer Nature 2019
H. Veendrick, *Bits on Chips*, https://doi.org/10.1007/978-3-319-76096-4_12

become impossible. *Design for testability* (*DFT*) and debug are commonly adopted as a general design approach, to enhance controllability and observability during the engineering phase of first silicon and to ease the detection of design bugs or physical failure mechanisms. Eventually, the results of all tests determine which of the chips pass and which fail.

The number of functional good dies per wafer, expressed as a percentage of the number of potential good dies per wafer, is called *yield*, which is also a function of how well the test covers the required functionality. This *test coverage* will never reach the 100% level. A flavour of the most important yield topics is therefore included in the discussions as well. Also a simple yield model illustrates how design margins and process defects impact the overall number of good dies on a wafer. Improving yield has direct influence on the cost of an IC. Design for manufacturability has therefore also become an integral part of IC design.

12.2   Testing

Testing is done to bridge the gap between customer requirements and quality of the design in combination with the manufacturing process. Testing thus helps to increase the quality of an IC. The yield is determined by testing and can be influenced by the complexity of the test: a simple test may lead to a higher yield but can lead to more customer returns. Large and complex ICs may have a relatively low yield, which will dominate the ultimate costs. How extensive a design needs to be tested depends on many different factors.

12.2.1   Test Categories

Three major test categories can be distinguished.

Characterisation tests are mostly executed manually and are developed for characterising the chip operating margins with respect to different operating conditions. These tests are more focussed on the accuracy of the test than on the speed of testing and are very much related to the parameters and type of circuits that are to be characterised:

- Design errors, design margins or manufacturing defects.
- The on-chip circuits: pure digital, pure analog, RF, pure memory or mixed signal.
- An increasing number of chip failures is related to dynamic effects such as interference, critical timing and noise. Most automatically generated tests detect the so-called *stuck-at-one* and *stuck-at-zero* faults, which cause circuit nodes to remain at '1' and '0', respectively, due to shorts or opens. Nanometre CMOS technologies, however, result in lower supply voltages and, consequently,

reduced noise margins. This will produce faults that are much more difficult to classify than the traditional stuck-at faults.

Production tests, which are performed in an automated mode, provide a way to reject those ICs that do not meet the required specification criteria or performance limits. Production tests include a large number of different tests to achieve the best possible test coverage and depend on the quality requirements of the target product and/or target application area:

- Consumer
- Communication
- Computer
- Aviation: aircraft and spacecraft
- Automotive
- Medical

Due to safety requirements, it will be clear that the latter three require higher test coverage.

Reliability tests, which are mostly performed manually, challenge the chip operation during and after the exposure to extreme electrical and environmental conditions:

- Electrical stress, burn-in.
- Temperature cycles, thermal shock and high-temperature storage.
- Increased humidity levels.
- Mechanical vibrations and shocks.
- Other reliability tests include electrostatic discharge (ESD) and latch-up, which are very much design related.

Because of safety requirements, aviation, navigation and medical applications usually require very exhaustive testing particular with respect to reliability standards. For the same reason, also automotive products require exhaustive testing to guarantee the safety requirements. Moreover, they operate in more 'hostile' environments, which may include large supply transients or interference caused by switching of heavy or inductive loads such as lamps and starter motors, high humidity, extremely low and high temperatures, etc. These extreme conditions require specific protection and more stringent reliability tests.

When all tests are executed properly, only a few of the chips that pass all tests still may be returned by the customer (*customer returns*; *escapes* because of a failure, which displayed itself only in the application), either directly or after a while (day, week, month, year). The number of *customer returns* is expressed in *ppm* (*parts per million*), which represents the ratio of customer returns per million supplied chips. This ppm level has become representative for the quality of a delivered product.

Ppm acceptance levels are related to the quality requirements of the application domain. While typical automotive applications allow 1 ppm, consumer applications and microprocessors may show ppm levels of 50 and 300, respectively.

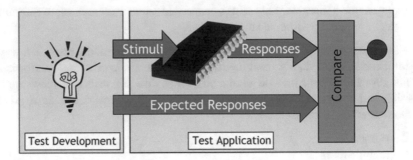

Fig. 12.1 Basic principle of IC testing

Tests can be performed both on the wafer and on the final packaged product. Pretests, also known as *e-sort* (early sort), are usually performed directly on the dies on the wafer to prevent unnecessary assembly costs. The final tests are performed on the packaged die. There is often a lot of overlap between the pretests and the final tests. As a result of the associated additional costs, the number of redundant tests must be limited.

During pretest, the individual ICs are tested on the wafer by probing the bond pads of the chip. A photograph of a *probe card* with more than 120 probes has already been shown in Part I, Fig. 4.8. A probe station brings these small needles into contact with the IC's bond pads. A test system provides predetermined stimuli for the IC and compares actual output signals to expected responses (Fig. 12.1). The stimuli should ensure that a large percentage of possible faults will result in discrepancies. This percentage is called *fault coverage* with respect to the applied fault model and is most commonly targeted at above 99%.

The test stimuli and response signals are transferred through a connector that provides a bidirectional link between the probes and the test system. The test system, *Automatic Test Equipment* (*ATE*), can also be used to control the wafer prober or handler to automatically step from one die to another so that a number of dies can be tested in rapid sequence.

It was relatively easy to manually determine test stimuli (vectors) for complete SSI (small-scale integration) circuits. For VLSI circuits, however, this is impractical and has led to the development of computer programs that generate test vectors automatically. A complete test program may consist of several subtests. The first test is the *contact test*, in which it is verified whether all pins of the probe card make contact with the respective bond pads. If so, *functional tests* are applied to check the correctness of the on-chip functional and memory blocks. Last but not least, a *supply current test* is done. This test evaluates the power consumption of the chip in standby mode to detect possible defects or opens, which may cause relatively large supply currents even when the chip is in standby mode and no signals are switching.

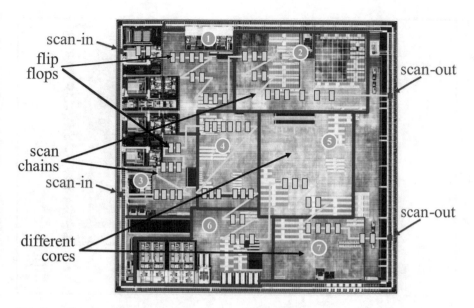

scan-in

flip
flops

scan-out

scan
chains

scan-in

scan-out

different
cores

Fig. 12.2 Example of the use of scan chains in a complex processor

The quest for higher bit and gate densities takes much design effort aimed at the realisation of a maximum amount of electronics on a minimum area. However, designers must ensure that their circuits remain testable.

12.2.2 Design for Testability

The previous sections presented an overview of the most important tests that are currently applied to test an integrated circuit.

It is therefore required that these tests are supported by the design. This is generally referred to as *design for testability* (*DFT*). In other words, several additional test circuits must be implemented to enable the various tests and to shorten the test time.

Because of the complexity of modern ICs, most on-chip functional blocks are not directly accessible through the I/O pins. In order to test their functional correctness, they all include so-called scan chains. These scan chains are formed by the already existing flip-flops in these blocks. These flip-flops have an additional input and selector circuit that enables them to bypass the logic gates and connect them directly to the output of the previous flip-flop to form a serial *scan chain*. Figure 12.2 shows a chip with various scan chains.

During the scan test of a certain functional logic block, the logic of the other blocks is bypassed, and their flip-flops are then only used as scan chains to transfer the test patterns from the I/O pads to the tested block and then from the tested block to other I/O pads. For instance, if we want to test block 4 in the above figure, we

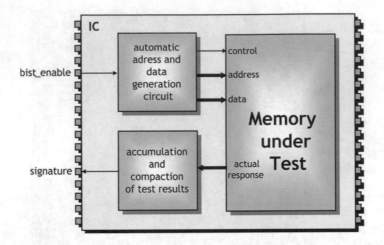

Fig. 12.3 Example of built-in self-test in an embedded memory

scan the input signals for this block through the scan chain of block 3. Once they have arrived at the inputs of block 4, the scan is stopped. Next, block 4 is put in the functional mode. After completion of its function, the chip is put back into the scan mode, and the output signals of block 4 are scanned through a scan chain formed by flip-flops in blocks 6, 5 and 7, respectively, to the lower right scan-out pin.

Therefore, a chip also requires additional test control circuitry to perform the overall scan test. Advanced chips can have many scan paths (e.g. 50) operated in parallel to shorten test time.

12.2.2.1   Built-In Self-Test (BIST)

The cost of testing will dramatically increase as a result of the increase in the number of functions on the chip, of the speed increase of the circuits, of the reduction of the voltages (smaller noise margins) and of the increase in the number of bond pads. The cost of a tester ranges from a few million to more several tens of millions US$ depending on the chip complexity. Therefore built-in self-test techniques can help to limit or reduce the overall test cost. *Built-in self-test* (*BIST*) techniques are currently used in many (embedded) memories. Figure 12.3 shows an example of BIST in an embedded memory: *memory BIST*. To reduce cost of overall chip testing, BIST techniques are also evaluated for inclusion in the design of digital and analog blocks.

Although the embedded memories in a VLSI chip contain most of its transistors, it is relatively easy to achieve large memory test coverage, because of the regular architecture of its memory array. To be able to detect defects between neighbouring bit lines or word lines, they have to be set in different logic states. First the complete memory is loaded with '1's and then read. Next it is loaded with '0's and then read.

Then a *checkerboard pattern* (1 0 1 0 1 0 1) is loaded into the memory, such that every '1' is surrounded by '0's and then read. Next the inverse checkerboard pattern is loaded and read. These tests only contain very regular patterns of '1's and '0's, which can easily be generated by an on-chip circuit, which also generates the addresses of the memory cells to write the '1's and '0's in. All output data (response data) is then accumulated and compacted in another circuit that generates a signature, which is dependent on all bits that are fed into it. If one or more bits are wrong, the signature does not match its expected value, meaning that it has detected a fault.

Because of its simplicity in both the on-chip generation and comparison of the test vectors, memory BIST has already been incorporated in many IC designs and has become more or less standard practice, today.

Currently, BIST is often combined with a repair action. This technique is called *built-in self-test and repair* (*BISTAR*) in which faulty columns are replaced by correctly operating redundant columns, by blowing polysilicon fuses in their address selection circuitry.

Introducing self-test into logic cores is much more complicated than in memories. This so-called *logic BIST* (*LBIST*) measures the response to random test patterns, fed to the different scan chains in the logic core. There are a few remarks to be made here. A disadvantage of LBIST is the associated low fault coverage. However, certain LBIST may show reasonable test coverage, but then it is at the cost of a relatively large area overhead. LBIST has therefore not yet become a mainstream test solution. It was and still is not an integral part of the design tools and design flow. However, with the rapidly growing test cost, LBIST may become more generally accepted as a standard design for testability methodology.

LBIST is already in use for some time in special applications, such as security, where a scan test would enable unwanted read out of the security key. Also aviation and certain automotive applications, which require repeated tests to maintain safety during their lifetime, use several forms of LBIST. These are a few specific examples of applications that require field tests and where there is no tester nearby.

12.2.3 Testing of Mixed-Signal Circuits

Most SoCs, today, contain a large mixture of analog, digital, memory and RF circuits. Testing of analog circuits is much more complex than that of digital and memory ones. An analog circuit is usually tested by verifying that its functionality meets its specifications. The input signals can easily be generated for circuits with a relatively small number of devices with only a few input and output signals. Many analog circuits are close to the periphery of a mixed-signal chip, because they directly interface with I/Os. However, the increasing complexity of mixed-signal circuits has dramatically reduced the direct accessibility potentials of the analog chip portions leading to a reduction of their observability and controllability as well. Moreover analog circuits have a large number of specifications, which makes a complete specification test very time-consuming and expensive.

Test methodologies for analog circuits are still relatively underdeveloped due to the complex nature of analog signals. There hardly exist commonly accepted analog fault models. Analog signals have an infinite number of states and are more sensitive to substrate noise and crosstalk. They are also more sensitive to process parameter variations. Generally, the testing of analog circuits is as difficult as their design. Therefore, the development of analog tests not only depends on the circuit functionality, but it is also very much related to the design/test engineers experience.

Efforts to reduce test complexity and costs focus on optimising (reducing) the number of specification tests without too much degradation of the test coverage.

More on testing of mixed signals can be found on the Internet [43, 44] but is beyond the scope of this book.

The most important message, to close this section on testing with, is that the test costs are currently so high that it is worthwhile to devote a few percent extra silicon area to support the testability in order to reduce the test cost with a much higher percentage.

12.3   Yield

The wafer diameter used in modern IC production is mostly 8 or 12 inches. The size of an IC determines the number of dies on a wafer. Most die sizes range between $20 \, mm^2$ and $200 \, mm^2$, and their number per wafer therefore ranges from a few thousand to several hundreds, respectively.

As discussed before, not every die on a wafer will be fully functional. Failing dies may be caused by systematic or random defects during the fabrication process, by critical mask dimensions, by design errors or by limited or too small design margins. The number of *functionally good dies per wafer* (*FGDW*), expressed as a percentage of the number of *potential good dies per wafer* (*PGDW*) is called *yield*.

The ultimate price of an IC is therefore directly related to the yield. Quite a lot of dies on a wafer do not meet their specified requirements during testing, particularly when the fabrication process is not mature. An additional number of dies is lost during packaging.

The yield observed during wafer probing depends on the production environment, the quality of the fabrication process and on the sensitivity of the design to process-induced defects.

12.3.1   Influence of Production Environment

The production of nanometre CMOS ICs puts very high demands on the factory building, the production environment, the production tools and the chemicals. Disturbances in the production environment may be attributed to the following parameters:

- *Temperature*: Fluctuations in temperature may cause the projected image of the mask on the wafer to exceed the required tolerances. Also several processing steps are done at elevated temperatures.
- *Humidity*: Humidity may cause a poor bond between the photoresist layer and wafer. This may result in under-etching during the subsequent processing step.
- *Vibrations*: Vibrations that occur during a photolithographic step may lead to inaccurate pattern images on the wafer and result in open or short circuits.
- *Light*: The photolithographic process is sensitive to UV light. Light filters are therefore used to protect wafers during photolithographic steps. The photolithographic environment in the clean room was often called the *yellow room* because of the specially coated lamps or filters used in it.
- *Process-induced particles and/or dust particles*: Particles that contaminate the wafer during a processing step may disturb the photolithographic step and damage the processed layer. This can eventually lead to incorrect circuit performance or to complete operation failures. For this reason, manufacturing areas are currently qualified by the class of their *clean room*. Modern advanced clean rooms are of class 1. This means that, on average, each cubic foot ($\approx$28 litres) of air contains no more than one dust particle with a diameter greater than 0.1 µm. In contrast, a cubic foot of open air contains 10^9–10^{10} dust particles that are at least 0.1 µm in diameter. The standard applied in conventional clean rooms required a class 1 room to have no more than one dust particle with a diameter greater than 0.5 µm per cubic foot. This was because smaller particles could not be detected. A conventional class 1 clean room is comparable to class 100 in the currently used classification. A lot of effort is invested in keeping the contamination level as low as possible. Clean room operators need to wear special suits to maintain high quality standards of the clean room with respect to contamination.

Silicon wafers are subjected to many process steps to build a complete chip. For advanced processes, today (2018), this number exceeds 1000! Each step requires physical treatment performed with a dedicated tool. Feature size reduction has constantly increased the requirements with respect to the purity of the chemicals, gases and environments that contact the wafers during processing. The exposure of the wafer surface to the less pure clean room environment introduces defects and results in yield loss. Modern clean rooms have class 10–100 for the overall environment. A mini environment, with controlled airflow, pressure and much less particles (e.g. better than class 1), is used to transport the wafer to the various process tools in the clean room. Such a mini environment is called a standard mechanical interface environment, a SMIF environment or *SMIF pod*. It protects the wafers from particle contamination and provides an automated and standardised interface to the process tools, which are also of class 1 or better. The wafers remain either in the SMIF pod or in the tool and are no longer exposed to the surrounding clean room airflow. SMIF pods are usually used for wafer sizes up to 200 mm. The front opening unified pod or *FOUP* mini environment (Fig. 12.4) was particularly developed for the constraints of the 300 mm generation. FOUPs have RF-identification tags for automatic handling of wafer batches in the clean room.

Fig. 12.4 Example of the use of a FOUP mini environment in a modern clean room. (Source: Entegris)

- *Electrostatic Charge*: Electrostatic charge attracts small dust particles. Very high charge accumulation may occur at low humidity. This can lead to a discharge which damages the electronic circuits on the chip.
- *The Purity of the Chemicals*: The used chemicals must be extremely pure to guarantee the high grade of reproducibility and reliability required for ICs.

12.3.2 A Simple Yield Model and Yield Control

The above parameters, the complexity of the process and the size of an IC determine the yield. Disturbances anywhere during wafer processing may cause defects. In order to control the production costs and predict the product's performance, yield loss mechanisms must be very well understood and accurately modelled. The basic cause of yield loss can be threefold. *Systematic yield loss* is usually caused by the sensitivity of the design to process variations or by the sensitivity of the lithographic or process steps to certain pattern topographies in the layout. These systematic failures are usually spatially or temporally correlated. *Parametric yield loss* is often caused by marginal operation of the design, e.g. critical timing, too much noise or too small noise margins. Finally, *random yield loss* is typically associated with physical mechanisms, such as metal shorts due to defects (particles) and contaminants or open contacts and vias due to misalignment or formation defects. These are usually characterised by the absence of any kind of correlation.

Fig. 12.5 Useful wafer area for PGDW

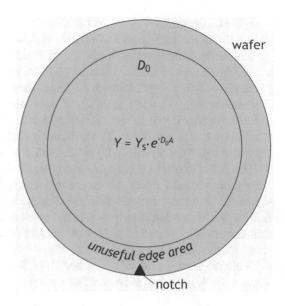

There exists several *yield models* today [45, 46]. Each model assumes a particular defect density distribution: exponential in the Seeds model, triangular in the Murphy model, gamma in the Negative Binomial model and random in the Poisson model, to name a few. IC producers compare for a specific process, yield data versus die size with results from the selected model to achieve the best fit.

For educational reasons, we will use the Poisson model, in which the overall die yield is described as a product of parametric/systematic yield Y_s and random-defect yield Y_r. According to this model, the yield Y is expressed as:

$$Y = Y_s \cdot Y_r = Y_s \cdot e^{-D_0 A} \tag{12.1}$$

where Y represents the pretest yield; D_0 the defect density (#defects/cm^2) during fabrication, which also depends on the product defect susceptibility; and A the die area. The influence of the die area on the yield is twofold. A smaller die not only results in an exponentially higher yield, it also results in an increasing number of dies on the wafer.

The parametric/systematic limited yield Y_s does not include the unuseful wafer area close to the wafer edge. The useful wafer area (see Fig. 12.5) is defined by the total area occupied by complete dies, with the exclusion of a narrow circular edge area (with a width of several millimetres) and a bottom flat side. Today's production lines use electrostatic clamping devices for wafer handling, which offers significant advantages over the conventional mechanical clamp ring by increasing wafer edge utilisation and yield. In current wafers (8 inch wafers and larger), the flat side is replaced by a *notch*, which is a kind of mark (incision) to convey wafer orientation for processing. The total number of dies within the total usable wafer area is called *potential good dies per wafer* (*PGDW*).

Y_s is also determined by the match between product design and process window. Especially in the early phase of process development, yield loss is dominated by parametric/systematic issues. Such defects are the result of structural failure mechanisms, which may be caused either by physical process defects or by an incorrect or process sensitive design. Both are relatively easy to find.

Random-defect limited yield Y_r is determined by non-uniformly distributed defects, which are mostly caused during processing steps that involve masks with very dense and fine patterns. These masks include those used to define patterns in transistor regions, polysilicon layers and in the lower metal, contact and via layers.

The defect density D_0 in Eq. 12.1 represents the density of defects causing uniformly distributed failures. These are uncorrelated and randomly distributed over the wafer. Examples include dust particles during lithography, which may potentially affect each mask exposure step, but particularly during exposure of the masks that contain high pattern densities. A process with smaller feature sizes becomes more sensitive to smaller sizes of defects.

Yield is directly related to cost per chip. The more *functionally good dies per wafer* (*FGDW*), the lower the cost. This number of *FGDW* is defined as:

$$FGDW = PGDW \cdot Y \tag{12.2}$$

The eventual production cost of a chip is determined by the cost of a fully processed wafer (*wafer cost*) and *FGDW*:

$$Cost / chip = wafer\ cost / FGDW \tag{12.3}$$

Clearly, the best way to reduce the chip *fabrication cost* is to increase the yield.

Particularly in the early phase of process development, Y_s will be relatively low, and D_0 will be relatively high. Figure 12.6 shows an example of the yield Y according to Eq. 12.1 as a function of the die area A for three cases for a 28 nm CMOS process.

Case 1 shows the situation during an early development stage of a new process, when $Y_s = 0.6$ and $D_0 = 2$ [defects/cm^2].

Case 2 may represent the situation after 6–10 months: $Y_s = 0.85$ and $D_0 = 0.5$ [defects/cm^2]).

Case 3 represents a more mature process: $Y_s = 0.97$ and $D_0 = 0.25$ [defects/cm^2].

This is the reason why many companies offer scaled versions of an IC relatively shortly after their original designs. The area will be smaller, and the process will be more mature, then. A process with smaller feature sizes becomes sensitive to smaller sizes of defects.

Yield is measured in various ways. Mostly it is related to the percentage of fully functional dies. However, since a complete chip test consists of many subtests, the tests can also differentiate between which of the functions or tests is/are failing.

Traditionally, during wafer test, an ink dot was deposited on every die that failed the test. Today, the distribution of correct and failing dies across the wafer, called

Fig. 12.6 Yield curves at
different stages of process
maturity

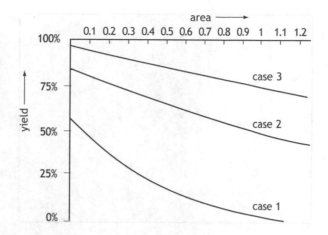

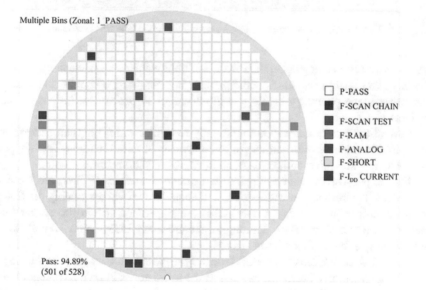

Fig. 12.7 Example of a wafer bin map showing correct and failing dies. (Source: NXP Semiconductors)

wafer map or *wafer bin map* (Fig. 12.7), is stored in the tester's memory. The test results are put in bins, presented by colour-coded dies on the wafer map. The tested chip is a mixed-signal design containing logic and memory blocks and analog circuits. All white dies in the figure pass all tests. The red dies fail the scan-continuity test, which verifies correct behaviour of all scan chains. The pink dies fail the full scan test of all logic cores on the die. The light-blue dies have at least one failure in the SRAM memory. The die in the second row from the top has a failure in one of the analog circuits. Finally the dark-blue dies fail the supply current test by a few factors, while the yellow-coloured die represents a full supply short. Wafer bin maps

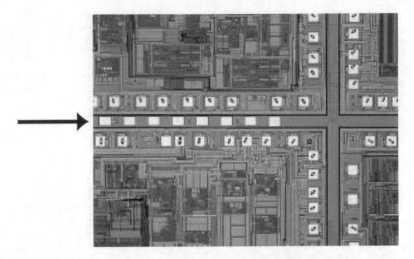

Fig. 12.8 Example of process control modules within the scribe lane between dies

can also be used to aggregate data from multiple wafers and stack them for cross wafer or lot (batch) analysis. Specific patterns in a bin map are usually an indication for process equipment problems, marginal design or process variations or mask defects.

Several tools exist that can automatically recognize wafer bin map patterns and can provide valuable information for the diagnosis of failure causes. This supports the designers and the foundries to ramp up yields in shorter time.

For the purpose of yield control, *process control modules* (*PCMs*) are included on wafers. Traditionally, a wafer contained about five PCMs reasonably distributed over its surface area. Today, these PCMs are positioned within the *scribe lanes* between the dies (Fig. 12.8). Although there are many PCMs on a 12 inch wafer, usually only a few of them are measured.

A PCM often contains transistors with varying widths and lengths for the electrical characterisation of parameters such as the transistor gain factor, threshold voltage, effective mobility, (parasitic) capacitances, etc. PCMs also usually contain relatively large structures that facilitate the measurement of possible shorts and opens through meander structures, for example. These structures are sampled during and at the completion of the chip fabrication.

Often, more than a hundred parameters can be measured on a PCM. During the introduction of a new process, the PCMs on all wafers may be measured. When a process becomes mature, usually a few wafers per lot and a few PCMs per wafer are measured. The measurement results are used as an early feedback to control the process.

It is clear that a high yield is essential for reducing the cost per chip. Memories, which have extremely high transistor and wire densities, are much more sensitive to defects, causing the possibility of occurrence of a defect to be much higher than in digital and analog cores. Therefore many memories include on-chip *redundancy* circuits. It means that the memory array is extended with a few additional redundant columns (or rows). During the memory test, the column addresses of the faulty cells are stored in the test computer. A laser is then used, which edits the corresponding bit line address which is then replaced by a bit line address of a redundant column, by blowing up fuses in the address selection circuits. When a faulty cell would be addressed for reading or writing, it would automatically select a cell in the corresponding redundant column.

Particularly in the early phase of the fabrication of a new memory product, when the process is not yet mature, the defect density would be high, and, without the use of redundancy, the *memory yield* would be extremely low. With redundancy, the yield can be improved with factors between three to more than ten times in this period.

Finally, when the correct dies are packaged, the final tests are done, which, besides functional and reliability tests, also check the connections between package and die. These final tests, in combination with the pretest (wafer test), must limit the number of customer returns to a minimum.

12.4   Design for Manufacturability

Over the last decade, *design costs* for an average complex ASIC have started to explode, from approximately $1 million in 1998 to several 100 million today. The total development of the most complex Intel and AMD and Apple processors is even in the order of several billion US dollars$. This, combined with reducing product life cycles and manufacturing yields, has increased the drive to reduce the number of *respins* and to ramp up the yield in shorter time, to meet time-to-market, quality and cost targets. Design rules form the real link between process technology and design. In conventional CMOS technologies, 'absolute' *design rules* (*DRC rules*) were sufficient to create circuits with relatively high yields. In current nanometre CMOS technologies (90 nm and beyond), these absolute design rules are no longer sufficient, and extensive yield evaluation must be performed before a design is sent to the fab. Particularly layouts are adapted to increase this yield. This so-called design for manufacturability (*DfM*) can reduce the design sensitivity to defects (opens or shorts), but it may also support the lithographic process. Additional *DfM rules* are required to make reliable designs, which are tolerant to photolithography and process deficiencies, in order to maintain a sufficiently high yield level. Figure 12.9 shows some examples of random failures.

There exists no uniform definition for DfM. Some include all effects that are potential candidates to reduce the yield: defects, shorts and opens, lithographic variations, process variations, power integrity, noise, electromigration, leakage currents, reducing noise margins, etc. Many of these effects may also affect the design robustness and product reliability [47]. Advanced design rule manuals may therefore contain between 500 and 1000 pages!

Fig. 12.9 Examples of random failures: particles causing a potential short. (Source: NXP Semiconductors)

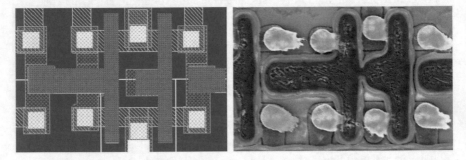

Fig. 12.10 Example of a systematic failure, showing the layout and a photograph of a polysilicon short. (Source: NXP Semiconductors)

DfM includes a set of guidelines and rules to make designs more robust against systematic, parametric and random yield loss and create more easy producible products. DfM is a way of anticipating on critical features or critical areas in the layout early in the design phase. Figure 12.10 shows an example of a systematic failure: a short between the two polysilicon areas in the right photograph. The photo has been taken after de-processing of the metal and via layers. Although the poly-silicon pattern fulfils the absolute layout design rules at this specific location, it still resulted in a short. It is probably the result of a combination of critical photolithography and etching steps.

Particularly at product introduction, when the design rules and process are not yet mature, the operating margins can be low and may cause parametric yield loss. Consequently, DfM rules may change as the process technology becomes more mature.

A few DfM rules have already become commonplace. Rules for wire widening and improved wire distribution (*wire spreading*; Fig. 12.11) were introduced around the turn of the millennium. Wire spreading is particularly an issue in those areas of the chip, where many wires are routed at minimum width and spacing, while there is ample room for wider wires at (much) larger-than-minimum spacing.

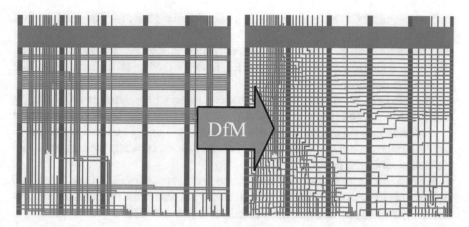

Fig. 12.11 Example of wire spreading to improve yield. (Source: NXP Semiconductors)

Today's ICs may have six to ten different layers of metal interconnections and hundreds of millions of contacts and vias. A contact is a small opening in the dielectric layer between the transistors terminals and the first metal layer. A via is a small opening in a dielectric layer between two successive metal layers. During the process these contact and via openings are filled with a conducting metal (tungsten or copper) to form the connections between the successive layers. These narrow contact and via connections have increased the risk of yield loss because they may fail to make a good electrical contact. These failing contacts or vias are called *opens*. Over the last couple of years, via *doubling* has entered the scene to reduce the number of opens on a chip. Doubling every via in a design is not possible, since it would have a serious area impact. The current approach is to double only the vias that have sufficient white space around them and which causes no area increase.

These additional DfM rules require the development of intelligent tools. These tools are usually applied during *chip finishing*, where they automatically perform the wire spreading and via doubling operation. An important requirement for applying these tools to achieve maximum yield improvements is that they are supported with adequate yield models that have the potential to recognize and fix critical layout areas without area increase. Including all forementioned DfM rules, a design rule manual, today, may have between 500 and 1000 pages!

12.5 Conclusion

The general requirement to create high fault coverage during the test of an IC is being challenged by an ever-increasing chip complexity. Advanced test methods have been developed to maintain high fault coverage, both during IC and board testing. Additional hardware is included in the design to support these methods. This also reduces test time, which could otherwise be a relatively large contribution to the ultimate price of an IC.

The purpose of IC testing is not only to separate the good dies from the bad dies, but the test results can also be used as feedback on the operating margins that the chip has with respect to its specification. Testing is also closely related to yield. Therefore a single yield model served to present a flavour of the most important aspects that determine the yield.

The hundreds of millions of transistors, combined with the required high performance, make testability and yield two aspects that have a huge impact on the overall cost of an integrated circuit. Therefore design for testability and design for manufacturability have become standard design tasks that are currently well supported by the design tools and flow.

When a test fails, it is not always trivial to find the cause of the failure. Therefore failure analysis techniques have become an integral part of the overall chip verification and analysis process, as will be discussed in the next chapter.

Chapter 13
Failure Analysis

13.1 Introduction

Not every chip operates fully according to the specification. Very complex ICs contain hundreds of millions to several billions of transistors and can have several hundreds to more than a thousand bond pads. It is therefore a tough job to locate the failure somewhere inside the chip, when, for instance, one output signal fails. The relation between an incorrect signal on one of the output pins and the location of an internal failure is very vague. As discussed in the previous chapter, dedicated advanced testing techniques are already included in the design to support testing. In many cases, however, this is not enough. Logical (design) errors are relatively easy to detect and to locate during scan test and in full functional test. The cause of timing errors, e.g. due to the occurrence of noise, can be so complex that even with the most advanced failure analysis techniques, it can take months before the correct diagnosis can be made.

This section gives an impression of conventional and advanced failure analysis techniques in order to explain the increased complexity of diagnosing the failure. Once the diagnosis has been made, the chip can be repaired directly by circuit-editing techniques to reduce both the number of respins and time to market.

13.2 Traditional Debug, Diagnosis and Failure Analysis Techniques

This section discusses failure analysis (FA) techniques that were already in place during the last century. They may still be in use, either for designs in conventional CMOS technologies with only a few metal layers or for designs in which special arrangements have been made. Critical nodes may have been taken up to the top metal, in the design, to enable probing of the chip from the frontside. Also the chip may be partly de-processed to create access to the potential failure location.

© Springer International Publishing AG, part of Springer Nature 2019
H. Veendrick, *Bits on Chips*, https://doi.org/10.1007/978-3-319-76096-4_13

Fig. 13.1 Picoprobes were used to measure a chip's internal signals. (Photo: NXP Semiconductors)

13.2.1   Diagnosis Via Probing

Probing is a method that allows us to measure any node that is available at the top level metal when there is still no passivation layer (scratch protection) on the wafer or when this layer has been removed locally. Conventionally, picoprobes were used (Fig. 13.1). They consist of needles as thick as a hair and with a very thin tip, less than a micron. This needle is connected to a very sensitive amplifier so that it can be used to probe very small signals on a chip. This technique was a reliable method for analysing incorrectly operating VLSI chips in semiconductor technologies with up to three metal layers. However, with the grow in the number of metal layers, it is becoming increasingly difficult to physically probe a signal that is only available in the one of the lowest metal layers. During the design phase, additional metal stacks could be placed at the critical nodes to create probe pads in the top metal layer. However, problems usually show up on locations in the circuit that were not expected to be critical. This means that the majority of failures can no longer be probed by these picoprobes. Alternative technologies have been developed to create access to the failing circuits.

13.2.2   Diagnosis by Photon Emission Microscopy (PEM)

Usually, when faults are detected on a chip, they almost always introduce a much larger local current, often causing a hot spot on a chip. When the electrons, which cause this current, decay to a lower state of energy, their energy surplus is converted

Fig. 13.2 Photoemission image of operating ARM microcontroller core in 90 nm CMOS, taken through the backside of the silicon. The 16 strong emission points in the core are part of the clock tree. (Photo: NXP Semiconductors)

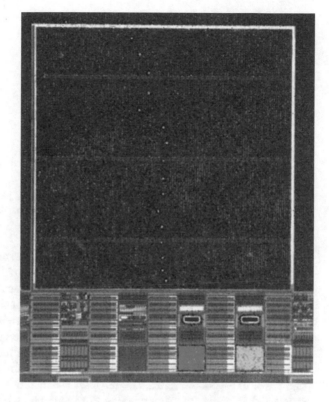

into photon emission (PE). With *photon emission microscopy* (*PEM*), the light that is emitted during chip operation is captured by a microscope and used for imaging. Figure 13.2 shows several hot spots on a CMOS chip.

Currents well below a microampere can be visualised at spatial resolutions of several hundred nanometres. In this way, latch-up, gate oxide defects, the switching of big transistors, shorts and/or opens, etc. can be detected. With the large numbers of metal layers as currently used in ICs, frontside PEM analysis faces severe limitations. Because silicon is transparent to near infrared light, infrared (IR) PEM is also used for die *backside analysis*, which means that the light that originates from a failure can be detected through the backside of the wafer.

13.3   More Recent Failure Analysis Techniques

The continuously growing complexity and density of integrated circuits, both in terms of number of transistors and timing requirements, have increased the variety of failure mechanisms. These failures can be originated by manufacturing defects or by design-related failure mechanisms.

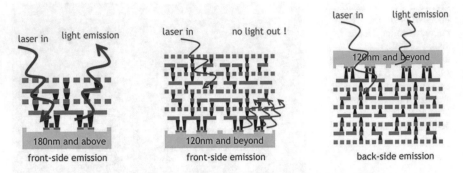

Fig. 13.3 Frontside FA techniques could be used in conventional technologies, with a maximum of three metal layers only, but with 120 nm CMOS and beyond, backside techniques are required. (Source: NXP Semiconductors)

Although everything is done to create robust and correct designs and optimise the fabrication process, still many ICs are not *first-time-right products* and require a lot of on-chip supporting circuitry to reduce the failure analysis and diagnosis time, in order to reduce the overall time-to-market. Several design measures are already taken to improve observability, both at system level as well as at the physical level. Developments in failure analysis techniques have enabled the way to access critical nodes from the backside of the wafer or chip, because of the above described inability to observe the transistors through analysis techniques from the frontside (Fig. 13.3). In products made in 180 nm CMOS and above, many failures could still be detected from the frontside with the PEM technique, because the resulting light emission could still propagate through the limited number (maximum three) of metal layers (left in the figure).

120 nm CMOS and beyond employ five or more metal layers, which form a barrier for the light emission through the frontside (middle). Therefore, many failure analysis techniques use *backside emission* to detect possible failures (right).

The following techniques are used to show the complexity and capabilities of state-of-the-art failure analysis techniques. One technique, which is based on photon emission microscopy but with improved sensitivity for backside usage, is the time-resolved photoemission microscopy. The others are based on stimulating the circuit with either laser beam or electron beam.

13.3.1   Time-Resolved Photoemission Microscopy (TR-PEM)

The basics of PEM have already been discussed before. During the switching of MOS transistors, light pulses are generated, as shown in the simple CMOS inverter of Fig. 13.4. When the input switches to high level, the lower nMOS transistor turns on, while the upper pMOS transistor turns off. The current through the nMOS will discharge the output (capacitance), causing a transition from a logic '1' to a logic '0' state. The current generates a small light pulse.

Fig. 13.4 The emission of a photon during switching transition of a logic gate

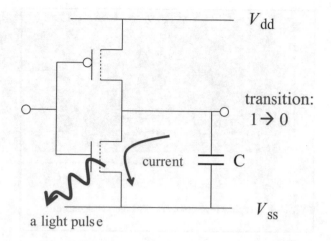

V_{dd}

transition:
$1 \rightarrow 0$

current

C

V_{ss}

a light pulse

When the substrate is thinned, a very sensitive, high-resolution, time-resolved camera can capture the integrated infrared component of these light pulses through the backside of the IC. The pulses are a measure for the switching moment and can be used to measure the exact moment of switching of any node in a digital circuit. Because not every switch generates a photon, it takes many hours to a day to aggregate sufficient photons for the creation of the switching-activity image in the scanned field. A TR-PEM tool is based on single-point detection which can capture single photons from a single switching node over a certain collection time and transports it via an optical fibre to the time-resolving equipment. It can create a measured waveform in a few minutes to an hour.

Figure 13.5 shows two such waveforms measured at different nodes on a chip with a time-resolved PEM. When one of these signals is the clock signal, this method can be used to analyse the timing behaviour of the suspicious signal(s).

The number of photons generated by a switch is much less than one and is dependent on the voltage swing. In a 90 nm CMOS technology, the number of photons per detected switch is in the order of 10^{-5}, which requires a relatively long time to collect sufficient photons to enable visualisation of the signal. For smaller technologies, with reduced supply voltages, the aggregation of sufficient photons has become much more time-consuming, making TR-PEM ineffective for sub-40 nm CMOS integrated circuits.

13.3.2   Laser Scanning Optical Microscopy (SOM) Techniques

There are different *scanning optical microscopy* (*SOM*) techniques. A scanning optical microscope may use an IC tester to create the required input stimuli and generate the optimum operating conditions to enable detection of even the smallest change in electrical performance. These conditions are usually such that the

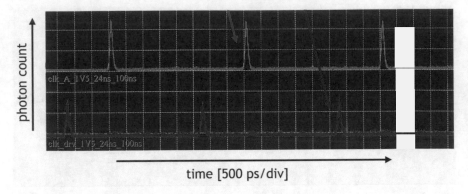

Fig. 13.5 A typical time-resolved PEM measurement result

operating point is set at the edge of correct operation. This point is usually derived from a shmoo plot measurement. Generally, a *shmoo plot* shows the operating area of a chip. In this particular shmoo plot (Fig. 13.6), the horizontal axis represents the frequency and the vertical axis the operating voltage. Each small square in the plot represents a chip measurement at a certain frequency and a certain voltage. The green area shows the combination of frequencies and voltages for which the chip operates correctly. In the red area, it does not. After the shmoo plot is completed, the chip voltage and frequency are set so that this chip operation point is exactly at the edge of the green area (e.g. the black circle in the left shmoo plot). In this point the chip is marginally operating, and the circuit that causes the failure will be very sensitive to any injection of heat and/or charge into its critical node. Next, the chip is scanned with a laser beam, which induces local thermal heating of the material (silicon, metal, etc.) in the laser spot or injects charge into a device. When the laser strikes the critical circuit that causes the problem, it will further slow down the circuit because of the local thermal heating by the laser spot. As a result of this, the operating area of the chip will become smaller as the green edge is moving upward, such that the chip no longer operates correctly in its original operating point (right shmoo plot in Fig. 13.6). Circuits in 45 nm CMOS and beyond may become faster when heated by the laser beam, and their starting operating point must then be chosen exactly in the nonoperating edge.

Many FA optical techniques can be used both from the top and backside of the chip, depending on what layer needs to be analysed. If the failure happens to be in one of the in-between metal layers, it becomes very difficult to create access to that layer. Optical backside analysis exploits the relative transparency of silicon to (near) infrared light. *Backside failure analysis* requires thinning and polishing of the substrate since the transmission of IR light through silicon decreases exponentially with its thickness. Particularly heavily doped substrates, which are much less transparent, often need to be mechanically ground down until a thickness in the order of 50 μm is reached and then further polished to achieve an adequate optical backside surface quality for proper light injection and propagation during laser-beam

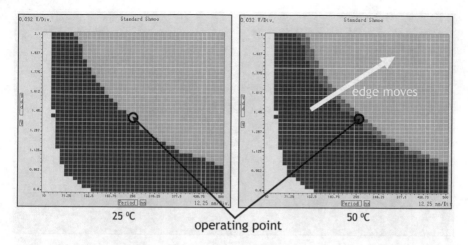

Fig. 13.6 Preferable operating point for failure analysis test

stimulation. Many CMOS circuits, today, employ lightly doped substrates (see Sect. 8.4.3) and do not require thinning at all. The package has a large influence on the ease of use of backside failure analysis. Needless to say, that backside analysis is easier for flip-chip packaged devices (see next chapter).

13.4 Observing the Failure

After the test has identified a failing chip, the debug and failure analysis techniques support the identification of the failure location. If the failure is caused by a manufacturing defect, such as a short circuit or an open circuit, it is necessary to make a detailed material analysis of the defect to understand its cause.

The next example shows the failure analysis process. It is related to a device, which has passed all tests, except for a speed test, meaning that the circuit does not operate correctly at the required clock frequency. By using a software tool that combines logged test data with an image of the die, the particular failing net could be localized. The result is depicted in Fig. 13.7, with the failing (white) net highlighted.

Next, backside SOM is used to visualise the response to the laser scan. Figure 13.8 (left) shows a zoom-in backside image of the failing net with an overlay of the layout to determine the position of this failing net in the circuit. Figure 13.8 (right) shows the complete highlighted failing net. No spots were visible at the driver side (top) of the net, but all three gates on the receiver side (bottom) of the net showed a response to the laser scan (green spots). This suggests that there exists a relatively high resistance somewhere in the interconnections between the driving and the receiving gates.

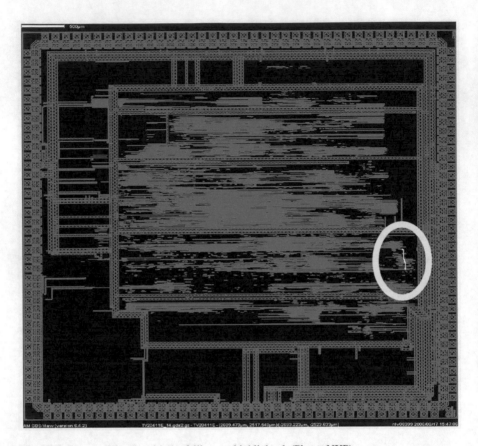

Fig. 13.7 Image of the die with the failing net highlighted. (Photo: NXP)

Because backside SOM did not show further responses to the laser scan, the device was opened from the top, and a frontside laser scan was performed. This yielded a clear response from only one VIA-5 between the fifth and sixth metal layer. Because this via was in the top via *layer* of the metal stack, it was not directly visible from the backside SOM analysis. Figure 13.9.a shows the failing via (blue spot around the via) on the frontside image of the IC.

The layout overlay helped to locate the coordinates of the defective VIA. Once the location is known, the failure analysis is continued to find the real cause of the failure. A thin cross section slice was cut from the chip by a focussed ion beam (FIB). Figure 13.9b shows the cross section as viewed through a transmission electron microscope (TEM). It clearly shows the defect in the VIA-5 layer, resulting from some etching problem in this particular via. Finally, this image showed that there was a bad contact between the VIA and the METAL. This caused the increased resistance in the delay path, which was the cause of the original speed test failure. This example shows that the complete failure analysis process, from a failing test,

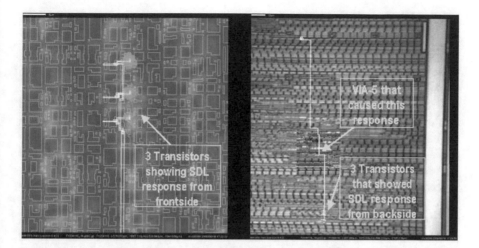

Fig. 13.8 Zoom in of a backside image with the SOM laser response and a layout overlay (left) and an overlay of failing track on a frontside image of the IC (right)

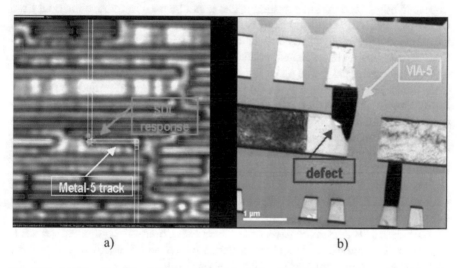

a)   b)

Fig. 13.9 (a) Zoom in of the SOM response of the failing VIA-5 in the overlay of the gds-2 file and (b) a cross-sectional image showing the defect. (Photo NXP)

down to the defect, is a costly, time-consuming and nontrivial effort, which may dramatically increase time-to-market of the final product. Therefore, a well-equipped failure analysis lab, combined with highly skilled FA engineers, is an essential part of the process of bringing a new design to volume production. Circuit-editing equipment, as discussed in the next paragraph, also belongs to the standard equipment of a failure analysis lab.

13.5   Circuit-Editing Techniques

Once the diagnosis has been made, the chip can be repaired directly by making and breaking techniques. The ability to physically edit an IC (circuit editing) may reduce the number of respins and helps to reduce time to market.

Traditionally laser beams were used for cutting lines as well as for laser-induced liquid-phase metal deposition to create interconnections on top of the scratch protection. A disadvantage of laser systems is that their resolution is limited, which has made them much less popular for *circuit-editing* techniques in deep submicron and nanometre ICs.

Focused ion beam (*FIB*) systems show better resolution, and, for cutting conductors, spatial resolutions of less than 2.5 nm have already been demonstrated. Operation of a FIB system is similar to that of a SEM. Instead of an electron beam, a FIB system uses a focused beam of gallium ions, which is scanned across the chip to image the sample in a vacuum chamber. At the location where the beam strikes the chip, ions and secondary electrons are emitted. The secondary electrons are captured, and their intensity is used to create an image of the surface of the chip. An important advantage of a FIB system is that it can be used to remove material from the surface of a chip (milling) or to deposit dielectric or metal layers (deposition). FIB is often used to physically edit a circuit on a chip. It can cut unwanted connections and deposit metal to change or add connections on top of the passivation layer or to create additional probe pads. Because holes can be made with high accuracy, even connections between different metal layers can be made, providing the capability to rewire ICs directly on the chip. These 'design modifications' may fix design errors or implement spec changes. Additionally, it enables the connection of an internal circuit node to a FIB-deposited metal area on top of the scratch protection. This will increase its load capacitance and can be used to correct timing violations. Figure 13.10 shows a schematic diagram of a basic FIB system.

It is equipped with a computer controlled gas injection system that can handle various different gases for the deposition of metal and dielectric material or for enhanced and selective etching.

A modern FIB system consists of complex and expensive equipment, which is capable of removing and depositing material (metal and dielectrics) and making smooth cross sections for SEM or TEM analysis [48].

It is sometimes combined with a SEM column for high-resolution imaging. To allow faster material removal at lower beam intensities, advanced FIB systems use gas-assisted etching. With this technique, holes can be etched down to the first metal layer. In this respect, holes with aspect ratios of up to 30 with a minimum feature size below 25 nm can be created.

The deposition of the conductive material on top of the scratch protection is easy but time-consuming. The turnaround time of the modified chips is in the order of several hours. The combination of a new mask and fabrication respin is very expensive, takes several months and introduces additional risks, since it is only based on simulations. A FIB circuit change allows the customer to perform all system-level

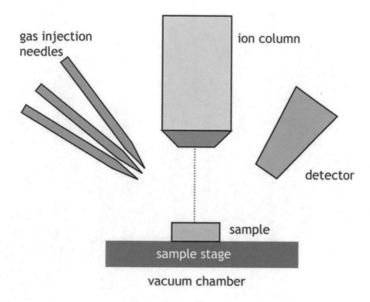

Fig. 13.10 Basic schematic diagram of a FIB system

tests and assures the next respin will include all the necessary changes. Since FIB technology also allows small holes to be accurately cut through the wafer, it may be a valuable tool in inspecting flip-chip packaging.

The large number of metal layers in today's ICs makes access to the low layers from the top very difficult. Because today's FIB machines also enable backside edit, wafers are backside ground such that the circuits can be accessed more efficiently from the backside [49].

13.6   Conclusion

The increased device complexity, combined with more levels of metal, reduces the fault observability. A design must therefore support both the test and the rapid iden-tification of the failure location and failure mechanism. When the cause of a failure cannot be traced with the combination of test/debug software and on-chip hardware, various techniques can be used to further analyse the failure. Because of the increas-ing complexity of integrated circuits, e.g. smaller feature sizes, increasing number of devices and metal layers and higher densities, semiconductor companies have installed very advanced failure analysis labs with complex tools. They support enhanced observability at both the front and backside and lead to shorter debug and failure analysis time, reduce cost and shorter time-to-market.

Chapter 14
Packaging

14.1 Introduction

The development of the IC package is a dynamic technology. Applications that were unattainable only a decade ago are now commonplace thanks to advances in package design. Moreover, the increasing demand for smaller, faster and cheaper products is forcing the packaging technology to keep pace with the progress in semiconductor technology.

The huge diversity of application areas, such as automotive, identification, mobile communications, medical, consumer and military, to name a few, combined with an exponentially growing device complexity and the continuous demand for increased performance has generated a real explosion of advanced packaging techniques.

Packaging is no longer a final step in the total development chain of a semiconductor product, and as such it has become an integral and differentiating part of the IC development and production process.

The *package* supports various important functions. It must:

- Allow an IC to be handled for printed circuit board (PCB; e.g. the motherboard in a PC) assembly and protect it during further PCB production.
- Provide both a mechanical and chemical protection against the environment.
- Provide a mechanical interface to the PCB.
- Provide a good electrical interface between PCB and chip.
- Enhance thermal properties to improve heat transport from chip to environment.
- Allow standardisation.

Currently, ICs may contain hundreds of millions to a few billion transistors. With such high integration densities, the IC package has become increasingly important in determining not only the size of the component but also its overall performance and price. Higher lead count, smaller pitch, minimum footprint area and reduced

© Springer International Publishing AG, part of Springer Nature 2019
H. Veendrick, *Bits on Chips*, https://doi.org/10.1007/978-3-319-76096-4_14

component volume all contribute to a more compact system implementation. Choosing the right package is essential in optimising chip and system performance and reduces costs.

14.2   Package Categories

Packages can be classified into different categories: board mounting techniques, construction form and power handling capabilities. The latter category offers high thermal capability, which enables IC usage in some of the most power demanding application areas.

Four major package categories that depend on different board mounting techniques can be distinguished:

- *Through-hole packages*, whereby the pins are inserted and soldered into plated through holes in a PCB. These packages are widely used in cheap electronic equipment where board space is less important. Example packages (Fig. 14.1) in this category are *single in-line package* (*SIL* or *SiP*), *dual in-line package* (*DIL* or *DIP*) and *pin grid array* (*PGA*).
- *Surface-mount packages* (dual/quad) are relatively thin packages (see *quad flat pack* (*QFP*) example in Fig. 14.1) whereby the leads are directly soldered on corresponding metal lands at the PCB surface. This allows smaller dimensions and components at both sides of the PCB.
- Surface-mount *area array packages*, which have an array of balls, or lands that are soldered on a PCB as well.
- Surface-mount *wafer-level packages* (bumped dies). In wafer-level packaging (*WLP*), most packaging steps are carried out directly at the wafer level instead of during the traditional assembly of each individual die after wafer dicing. Because the final package is almost of the same size as the chip itself, WLP is a true *chip-scale packaging* (*CSP*) technique. The IC production process is then extended and includes additional interconnection (redistribution) and protection steps.

The package choice depends on the expected priority in properties of the application, e.g. high density (very small package), high bandwidth (many pins and low self-inductance), high power (good thermal behaviour), etc.

SIL   DIL   PGA   QFP   MLF/QFN   FLIP-CHIP

Fig. 14.1 Various package images

14.3   Die Preparation and Packaging Process Flow

Before the actual packaging starts, each wafer will have to be ground down at the backside to the optimal thickness. After thinning, typical *wafer thicknesses* are between 250 μm and 400 μm, while 75 μm or less is more common for ultrathin electronic products.

Once ground to the required thickness, the wafer has to be separated into individual dies. These dies are locally separated from each other by narrow *scribe lanes* (see Fig. 12.8). There are several methods to separate a wafer into dies. As with glass, silicon can be scribed with a diamond tip and then broken along the scribe line by careful bending. An alternative was to cut the wafer by means of a diamond saw. Today, laser separation is more popular way of separating the dies. Laser dicing has a couple of advantages compared to diamond sawing. It is faster, it causes less material stress, it requires a smaller scribe lane and it is able to dice devices with different form factors on the same wafer.

After dicing, the real packaging process can start. The package choice is very much related to the electrical, thermal and size requirements dictated by the application domain. Two main interconnect technologies can be identified to realise the electrical connections to the die. The most common one is *wire bonding (WB)*, which is still responsible for more than 80% of all chip connections. Traditionally, gold and aluminium wires were used. Today, copper is rapidly gaining a foothold because it has many advantages over gold: cheaper, better electrical and thermal conductivity and increased reliability. However, copper easily oxidises at relatively low temperatures and therefore requires a more complex overall bonding process.

Before the actual wire bonding can take place, the die is first mounted on a carrier (lead frame or substrate) in a process called die attachment. During this die attachment, an adhesive is deposited on the carrier, and the functional good dies are picked from the wafer and placed in the adhesive (Fig. 14.2). Which dies are being picked is determined by the wafer map (see Fig. 12.7) that has been generated during probing. Adhesive materials are typically a mixture of epoxy and a metal (aluminium or silver) to ensure a low electrical and low thermal resistance between the die and package. For thermally enhanced applications also, solder can be used to attach the die to the carrier. IC reliability is strongly influenced by the quality of the bond wires. Diameters of the wires range from 15 μm for fine-pitch applications to 150 μm for high-power devices.

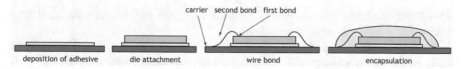

Fig. 14.2 Overview of the wire bonding process

Fig. 14.3 Examples of a double- and a triple-row wire bond interconnects. (Source: NXP Semiconductors)

During the bonding process (Fig. 14.2), the bonding tool is guided to the bond pad on the die. This so-called first bond is achieved by using thermal and ultrasonic energy. Next, the wire is stretched to the corresponding finger of the lead frame on the carrier, and again by using pressure and ultrasonic energy, the opposite end is welded to the lead finger to form a stitch bond (also known as wedge bond). At the formation of this second bond, the wire is also automatically cut in preparation for the next bond. After all the pads have been bonded, the die is encapsulated.

The number of transistors per unit area grows quadratically with the scaling factor. Because the pad positions are usually only at the chip periphery, its number only grows linearly with it. This has caused a lack of bond pad positions at the chip's periphery and drove the demand for multi-row bond pad (*staggered bond pads*) connections to support a variety of applications. Figure 14.3 shows examples of a double and a triple row of wire bond interconnections. It also shows the extreme requirements with respect to the high accuracy of the bonding tool.

In *flip-chip bonding* (*FCB*), which is the second interconnect technology, the die is assembled face down directly onto the circuit board (Fig. 14.1 right) with solder, gold or gold/nickel bumps. Compared to wire bond, this technology comes with less area overhead, because there is no additional area needed for contacts on the sides of a chip. It enables the final packaged chip to be only marginally larger than the original die (*chip-scale package*). In the example of the *controlled collapse chip connection* (*C4*) soldering process, first, solder bumps are deposited on the die bond pads (Fig. 14.5), usually when they are still on the wafer, and at the corresponding locations on the substrate. Figure 14.4 shows two rows of dies sawn from the wafer. The zoom-in shows a wafer-level CSP with the deposited solder bumps. It also shows that a *redistribution layer* is used.

During die placement (Fig. 14.5), the die is flipped upside down, and the array of balls on the die is aligned with the array on the substrate. Depending on the FCB technology, it is then either pressed or reflowed (melted), or the complete embodiment is reflowed (melted) in a furnace, to create all electrical connections. During this reflow step, the chip is self-aligned to its exact position on the substrate by cohesion. If no isolating material was deposited before, there will always be a gap

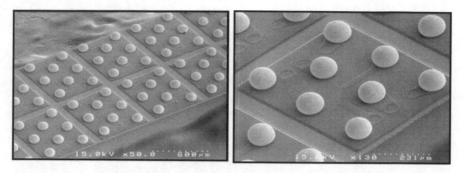

Fig. 14.4 A wafer-level chip-scale package with direct ball drop

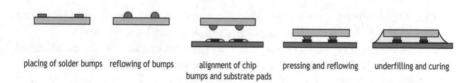

placing of solder bumps   reflowing of bumps   alignment of chip   pressing and reflowing   underfilling and curing
                                                      bumps and substrate pads

Fig. 14.5 Overview of flip-chip bonding process

between the die and substrate. In the next step, the die is encapsulated with good isolating material (epoxy) to fill this gap (underfill). This underfill serves to protect the chip from humidity and impurities but also improves reliability in terms of mechanical and thermal stress.

Finally the FCB process is completed by curing (hardening) the underfill material by using heat or light at a certain wavelength, depending on the underfill material (e.g. UV light for a UV cure material). Flip-chip bonding accommodates dies that may have several hundred bond pads placed anywhere on their top surface. In many cases an additional metal *redistribution layer* is required to transfer periphery wire bond pad connections into an area array of connections for flip-chip bonding. Compared to other wire bonding, FCB results in very short connections and exhibits improved performance in high-speed applications.

14.4   Electrical Aspects of Packaging

The drive for higher bandwidths and the resulting increase in signal speed, combined with the ever-present demand for area and cost reduction, has caused the package development to become a significant and integral part of the total development process of integrated circuits. Package costs are closely related to the number of package pins, by the size of the package and by its thermal resistance. A package with poorly controlled electrical parameters (resistance R, inductance L, capacitance C and impedance Z) needs more pins than a package that meets the electrical signal interface design criteria.

The electrical characteristics of a package are determined by its construction. The type of signal interface indicates the desired *RLCZ* of the package interconnects. A high-speed differential interface may require 100 Ω differential impedance between metal tracks, while minimisation of the power and ground pins' inductance is more important for a single-ended interface. The signal type will determine the desired electrical characteristics of the package interconnect.

Very high current variations (d*i*) during very short periods of time (d*t*) in high-speed ICs can cause an increased *voltage drop* Δ*V* across the bond wires. This is mainly due to the *inductance* (*L*) of the wires and is expressed as follows:

$$\Delta V = L \frac{di}{dt} \tag{14.1}$$

This voltage drop may become critical in analog/mixed-signal and high-speed digital circuits, unless suitable design measures and margins are taken.

Certain package types, however, can only support certain ranges of electrical parameters. Conventional package types with built-in lead frames have relatively large pin/lead inductances (2–50 nH), because of longer lead lengths. They also tend to have a high mutual coupling. Ceramic multilayer packages, such as pin grid arrays (PGAs), have better performance due to the presence of power and ground planes but are relatively expensive. As previously discussed, the use of flip-chip bonding can also improve electrical performance by minimising the lengths of the connections between the die and the substrate, resulting in inductances of 0.5–1 nH. Recent developments in package technology, however, allow more flexibility in the design of package interconnects to meet cost targets. Some package types allow the design of specific impedances, while others try to minimise them altogether. *Ball grid array* (*BGA*) packages use inexpensive laminate substrates allowing the inclusion of power and ground planes and therefore the design of transmission line structures. On the other hand, the so called no-lead packages, such as QFNs, have virtually no pins and sometimes not even bond wires, thereby minimising the influence of the package on the overall electrical behaviour of the chip. It is therefore extremely important to understand that the requirements of the interconnections are dictated by the application. Particularly in high-speed applications, a reasonably accurate electrical model of the package is an essential part of the final performance simulations of the integrated circuit.

In the packaging of ICs, we distinguish several hierarchy levels of interconnections:

- First level of interconnection: die-to-package connection.
- Second level of interconnection: package to PCB connection.
- Third level of interconnection: PCB wires.
- Fourth level of interconnection: PCB to system (backplanes) connection.

It should be clear that not only the die-to-package connections must be optimised for high-performance ICs; also the other levels of interconnections must be

Table 14.1 Thermal resistance values for different packages under still-air conditions

Package pins	Package designator	Package outline code	R_{JA} [°C/W]	R_{JC} [°C/W]
PLCC -20	A	SOT380	80	32
PLCC -52	A	SOT238	48	13
SSOP -20	DB	SOT339	136	40
SSOP -48	DL	SOT370	88	25
TSSOP -48	DGG	SOT362	104	23
QFP-52	BB	SOT379	62	15
VFBGA -56	EV	SOT702-1	80	21

Source: NXP Semiconductors

optimised and adapted to each other to achieve maximum overall system performance. This becomes even more important with the state-of-the-art packaging technologies, such as multi-chip modules, stacked dies, system-in-a-package (SiP) and system-on-a-package (SoP). These packaging technologies are discussed in Sect. 14.7.

14.5   Thermal Aspects of Packaging

Another dominating parameter in the performance and reliability of an integrated circuit is the physical temperature of the die inside the package, which is determined by the power consumption of the IC in combination with the thermal behaviour of the package. This requires a strong interaction between the IC, the package, the system design and its application. The most commonly used but simple model for IC packages includes two thermal resistance parameters. For a given power dissipation P, the junction-to-air *thermal resistance* R_{JA} represents the ability of a package to conduct heat from junction (die) to ambient and is expressed as follows:

$$R_{JA} = \frac{Temp_J - Temp_A}{P} \quad [°C/W] \tag{14.2}$$

where $Temp_J - Temp_A$ represents the temperature difference between the die (junction) and its environment (ambient). R_{JA} is often determined corresponding to the JEDEC [50] requirements for standard test boards and in different airflow conditions, including still air. Table 14.1 shows some values for R_{JA} under still-air conditions for different package types. More thermal resistance values for other types of packages can be found in [51].

In many applications the maximum junction temperature is defined as 125 °C. If we assume a consumer application with an ambient temperature of 70 °C, the maximum allowed power consumption of an IC, packaged with a 48 pins SSOP (see Table 14.1) under still-air conditions (on a reference board of the supplier), is then equal to:

$$P = \frac{Temp_J - Temp_A}{R_{JA}} = \frac{125 - 70}{88} = 625 \text{ mW}$$

If the power consumption is more than this calculated maximum, either a heat spreader is required or airflow must be introduced, using a fan.

The other parameter is defined as the junction-to-case thermal resistance R_{JC} and represents the ability of a package to conduct heat from the junction (die) to the surface (top or bottom) of the case (package) and is expressed as follows:

$$R_{JC} = \frac{Temp_J - Temp_C}{P} \quad [^\circ C/W] \tag{14.3}$$

This parameter is only applicable if an external heat sink is used and the heat is only conducted through that surface that connects to the heat sink.

If we assume an IC consuming 1 W, which exceeds the above-calculated maximum allowed power under the same conditions, then the required thermal resistance of the device must be equal to:

$$R_{JA} = \frac{Temp_J - Temp_A}{P} = \frac{125 - 70}{1} = 55 \ ^\circ C/W$$

This can either be achieved by introducing airflow or by using an external heat sink. This model only describes the steady-state heat conduction capability and does not account for the dynamics in power behaviour of the product in a real application. Heat flows are rarely one dimensional.

Different application boards or stacked packages change the environment of the product and can have a huge impact on its thermal behaviour. However, the value for R_{JA} can very well be used to compare thermal capabilities of different packages.

An accurate model for a particular thermal situation including two- and three-dimensional heat conduction paths may easily result in a very complex network. This has led to the development of compact thermal models, describing the thermal behaviour with an accuracy of 5% by using a thermal network with seven or more nodes connected by thermal resistances. A discussion of such compact models is beyond the scope of this book. An example is described in [52].

14.6   Quality and Reliability Aspects

Reliability is playing a role during the packaging of the die itself and also after the packaging. Various *quality* and *reliability tests* are applied to packaged ICs before they are approved for sale or used in applications with high-volume production. Many of these tests are standardised. An insight into the background to these tests and their implementations is provided in the next subsections.

14.6.1   Reliable Bonding

There are a few packaging aspects that are related to the reliability of the chip. First of all, the trend to reduce the dielectric constant of the inter-level dielectric (ILD) layers (*low-k dielectrics*) used during chip fabrication makes these dielectrics more porous, less robust and more sensitive to physical pressure during test (probing) and bonding. Secondly, in a *copper back-end* process, copper is used for all metal layers including the one(s) used to create the bond pads. However, copper oxidises quickly when subjected to air, and the oxidation prevents the creation of a good and reliable electrical contact between the bond wire and the copper pad. Therefore, during an additional re-metallisation step, a so-called *alucap* (aluminium cap) layer is formed above the copper pad area to create a good electrical contact with the bond wire. But this does not solve all reliability aspects. Particularly the drive for finer pad pitches and smaller pads requires probe cards with smaller and sharper needles, which increases the probability to punch through the alucap and expose the underlying copper. Also these exposed copper areas oxidise quickly, showing the same problems as described above. A solution to this problem is to increase the alucap area such that the probe needles do not land in the wire bond region (Fig. 14.6) and can no longer damage the underlying copper layer because it is separated by the passivation layer.

The increasing number of pads, combined with the drive for smaller chip areas, has forced the semiconductor industry to create *bond-over-active* (BOA) layout techniques, in which bond pads are not only located at the chip's periphery but also on top of active silicon areas at the periphery of the die core area on top of diodes,

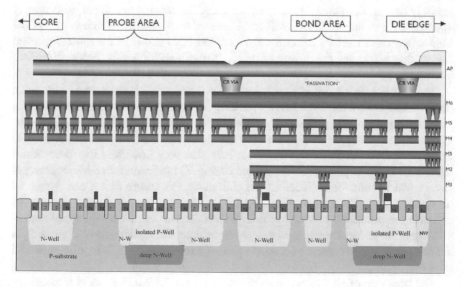

Fig. 14.6 Example of extended Al cap area to prevent pad damaging during probing. (Source: NXP Semiconductors)

power and ground lines, I/O transistors and ESD protection circuits. BOA may lead to a significant reduction in die size [53].

As mentioned before, traditionally, gold and aluminium wires were used, while copper is rapidly gaining a foothold. Copper wires have many advantages over gold wires: they are cheaper, have better electrical and thermal conductivity and increased reliability. However, copper easily oxidises at relatively low temperatures and therefore requires a more complex overall bonding process.

In summary, any change in pad-related design and IC process technology concepts may have severe consequences for the reliability of the bonding process and for an overall reliable chip operation.

14.6.2 Quality of Packaged Dies

Integrated circuits are exposed to many possible sources of damage, both during and after the manufacturing process.

One of the principal causes of damage is *electrostatic discharge (ESD)*, due to the transfer of charge between bodies at different electrical potentials. ESD pulse durations are very short and normally range from 1 to 200 ns, but they may introduce very large power spikes. The high impedance of MOS input circuits makes them particularly vulnerable to physical damage when they are exposed to these spikes. These may result from operations during the fabrication process, from handling (un)packaged dies or during bonding. It may also occur during testing and maintenance or in the application. While only a few devices or connections may be severely damaged, many more may suffer a level of damage that is not immediately apparent. These latent failures will result in *customer returns*, which is one of the biggest worries of semiconductor vendors. Thus ESD is one of the most important factors that determine the quality of an IC. It may also trigger the below-discussed latch-up.

The damage caused by ESD is irreversible. The human body is one of the main sources responsible for ESD. Just by walking on a carpet on a low-humidity day, for instance, a person, wearing shoes with highly insulating soles, can build up a voltage in excess of 30,000 V. The resulting charge can then be transferred via an ESD to an electronic circuit during touching. It is also very important that precautions need to be taken to prevent ESD damage during IC fabrication. In addition, protective measures must be included in the chip design to ensure that it can withstand acceptably large ESD pulses. On-chip MOS *ESD protection circuits* are used to increase the immunity of an IC to ESD pulses. These circuits are designed to provide input and output circuits with low-impedance shunt paths, most commonly a set of diodes, which prevent excessive voltages to arrive at the IC's input, output and core transistors.

The presence of nMOS and pMOS transistors in a CMOS process leads to the creation of parasitic bipolar npn and pnp transistors in all CMOS chips. Together, these devices form parasitic *thyristors* between the supply and ground terminals.

Activation of a thyristor results in *latch-up*. In this state, certain nodes on the chip will permanently be pulled low, while others are pulled high. The result is a dramatic increase in current consumption and a chip malfunction. In fact this erroneous state is latched, which can only be undone by disconnecting the chip or system from the power supply.

A chip's latch-up sensitivity can be tested by sequentially applying a voltage of one and a half times the maximum specified voltage to each pin while limiting the available current to 500 mA, for example. The actual chip current consumption is observed for signs of latch-up. The possibility of occurrence of latch-up can be reduced in the chip fabrication process by using low-resistive substrates (Sect. 8.4.3), or by reducing the supply voltage. In advanced CMOS chips, therefore, the latch-up sensitivity is low and likely to disappear inside electronic circuits, as the supply voltage, which is close to 1 V, is generally too low to trigger the parasitic thyristor. However, at the chip I/Os, the requirements on latch-up remain relatively high, since many applications still require a higher interface voltage (1.8 V, 2.5 V and 3.3 V).

Vulnerability to electrostatic discharge (ESD) and sensitivity to latch-up are two important quality criteria on which chips are tested.

14.6.3   Reliability of Packaged Dies

The increasing complexity of ICs means that their reliability has a considerable effect on the reliability of electronic end products in which they are applied. Reliability is therefore an important property of an IC and receives considerable attention from IC manufacturers.

Traditionally the failure rate of an electronic product is represented by the well-known *bathtub curve*, shown in Fig. 14.7. In the early life, the failure rate starts at a high level but is rapidly reducing as failing products are quickly detected and rejected and failure mechanisms due to immature production steps, material defects or design errors are investigated and eliminated.

During normal life, when the failure rate is constantly low, the product lifetime is expected to last till wear-out, due to ageing effects. Normal life failures are usually caused by random stress phenomena, such as the occurrence of a soft error. Wear-out failures become more likely when the product becomes older. Ageing is a physical process that causes deterioration, fatigue or depletion of material. The useful lifetime of a product is determined by the worst-case component, the one that lives shortest. Manufacturers must guarantee that their products function correctly over the intended lifetime. Therefore many different reliability tests have been developed to characterise the product quality to achieve a sufficiently long normal life for their products.

Related tests subject an IC in active and non-active states to various stress conditions. This facilitates rapid evaluation of the IC's sensitivity to external factors such

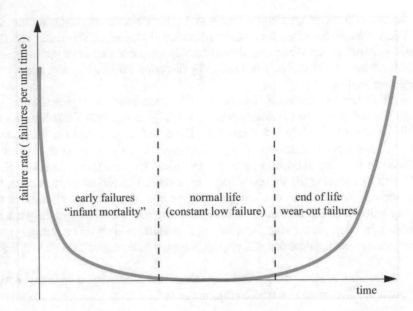

Fig. 14.7 Example of a bathtub curve, showing the failure rate of an electronic product over its life cycle

as temperature changes and humidity. The following discussions only provide a glimpse of the most important reliability tests:

Electrical Endurance Test This test exposes an IC to a high temperature (125–150 °C), while its supply voltage exceeds the specified maximum. Constant and varying signals are applied during the test, which may last for 1000 h. These stress conditions will make the weak devices fail, which is detected by applying normal tests that monitor whether the ICs still show correct functional behaviour.

The electrical endurance test reveals the following:

- *Infant mortality*, i.e. faults which are likely to arise in the early months of an IC's normal application.
- *Early failure rate*, i.e. faults which are likely to arise after half a year.
- *Intrinsic failure rate*, i.e. the probability of a fault occurring during a specified number of years.
- *Wear-out*, i.e. effects of prolonged use on the product.

Faults that are observed during the electrical endurance test can usually be traced to errors in the manufacturing process, which preceded IC packaging.

Temperature Cycle Test This test emulates practical temperature changes by exposing the product to rapid and extreme temperature variation cycles. The minimum temperature in each cycle is between −55 °C and +65 °C. The maximum temperature is 150 °C. The number of cycles used is typically 500. The test is carried out in an inert gas and/or an inert liquid. The main purpose of the temperature cycle test is to check the robustness of the package and the robustness of the connections between the package and its die. The test should reveal possible incompatibilities between the temperature expansion coefficients of the various parts of an IC, e.g. the die, the lead frame and the package material.

Humidity Test   This test exposes an IC to a relative humidity of 85% in a hot environment (85–135 °C). The test reveals the effects of corrosion on the package and provides an indication of the quality of the scratch-protection layer. Usually, the corrosion process is accelerated by applying different voltages to consecutive pairs of pins, with 0 V on one pin and V_{dd} on the other. Most humidity tests last 1000 h.

The required specifications of an IC depend on its application field, envisaged customer, status and supplier. It is clear that product areas, such as automotive and aviation, may require stronger reliability test conditions than consumer products.

It can take a relatively long time before the quality and reliability of a new IC in a new manufacturing process reach an acceptable level.

14.7   Trends in Packaging Technology

Figure 14.8 shows the trend in use of the various package categories, with a focus on density increase. Through-hole packages (e.g. TO3P and SIP) were the most preferred ones during the 1960s and 1970s, while surface-mount device (SMD) technology (e.g. TSOP, QFP, flip chip) became very popular during the 1980s and 1990s at the cost of the through-hole packages.

Products like mobile and smart phones get smaller and thinner every year, which automatically require the same shrink for the components they are built from. This means that conventional leaded parts, such as quad flat packs (QFPs), will increasingly be substituted by leadless parts like QFNs or even bare dies (wafer-level *chip-scale package* (*CSP*) (WLCSP)).

The complexity of nanometre ICs has reached an incredibly high level and will continue to increase. As discussed before, it puts severe demands to the density of die pads and package connections (pins or balls). This drives the trend towards area array packages, e.g. *ball grid arrays* (*BGAs*), in which solder balls form the connection between the package and the application board. In literature, BGAs are frequently combined with CSPs, which are usually referred to as packages whose sizes are less than 20% larger than the die itself. Most CSPs are wafer-level packages. The peripheral bond pads are then redistributed and rerouted to an area array of pads, using a thin film-like technology, which can be executed as an extension of the wafer fab process. An alternative is to send the wafer to a bump supplier who creates the redistribution directly on the wafer. Balls are placed on the rerouted pads by means of direct ball attach, to create the CSP (Fig. 14.9). Next, flip-chip bonding technology is applied, when the CSP needs to be attached to a kind of laminate carrier.

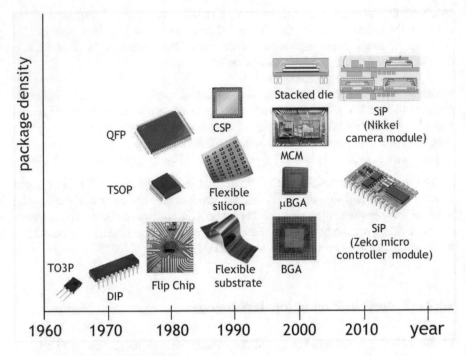

Fig. 14.8 Worldwide IC package density trends

Over the last couple of years, bulk CMOS processes have enabled the integration of digital, RF and mixed-signal functions onto one single die. Time has come to make a trade-off between traditional board design with discrete passive devices and a fully integrated system solution like a *system-on-a-chip* (*SoC*).

A SoC is an extremely integrated single-chip solution built from in-house and/or external IP. It contains the computing engine (e.g. microprocessor and/or DSP core), analog and digital cores and various memories and interfaces on the same chip.

The key benefits of a SoC realisation are:

- Better performance due to smaller on-chip signal delays, compared to chip-to-board signal delays, which are often an order of magnitude larger.
- Small physical size.
- Reduced overall system costs, due to a reduced number of components.
- Less power consumption.
- Increased reliability due to a reduced number of system components.

However, there are also some critical remarks to be made here. For many applications, the time between inception and high-volume production of a SoC may take several years. Many SoCs are therefore expensive, custom-designed products for high-volume market segments with relatively long lifetime expectancy. The increasing diversity of the system's applications requires the development of more sophisticated IP.

Fig. 14.9 Rerouted
wafer-level CSP

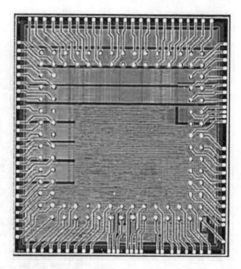

Today, system complexity is growing at a faster rate than that of a SoC and a printed circuit board (PCB). In many applications, Moore's law only deals with the integrated fraction of the system, leaving the largest part to relatively large discrete passive components such as displays, sensors, antennas, filters, capacitors, inductors, resistors and switches. In the example of a smart phone, less than one-fourth of the system consists of ICs, with the remaining part being passives, sensors, boards, interconnections and switches. Alternatives to SoC integration are the use of stacked dies or *system-in-a-package* (*SiP*) technology (Fig. 14.8). The latter usually refers to a single package that includes a multiple of interconnected integrated circuits and/or passive devices. SiP technology enables hybrid systems built from subfunctions that may have been created by different design methods in different technologies. Some people see a SiP and *multi-chip module* (*MCM*) as the same system solution. However, MCM is usually referred to as the integration of different dies on the same plane on the same substrate in one single package, while SiP also refers to stacked dies and/or passives in one single package. SiPs may use a combination of different packaging technologies including wire bond, flip-chip, wafer-level packages, CSPs, stacked dies and/or stacked packages. Figure 14.10 shows a wire-bonded MCM and wire-bonded stacked dies.

Compared to a SoC realisation, a SiP approach offers much more flexibility in adding new functions and features to the system. As is the case with all packaging technologies that combine one or more naked dies, also SiP technology faces the challenge of *known good die* (*KGD*), which is a chip that has been extensively tested before being placed into its package. When both an expensive processor and a cheap peripheral chip are to be combined onto a single substrate or into a single package, an almost 100% guarantee is required that this peripheral chip will operate fully according to its spec. This is to prevent to throw away the total substrate, including the expensive processor, if only the cheap peripheral chip does not work properly. To avoid this problem, a new upcoming trend can be identified: *package-on-a-*

Fig. 14.10 Example of an
MCM and of wire-bonded
stacked dies. (Source:
NXP)

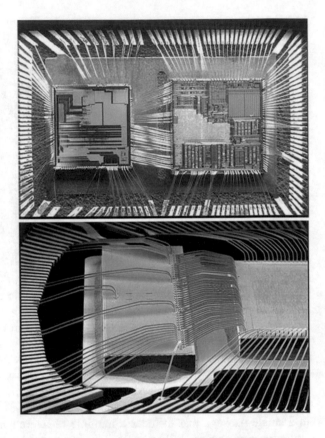

package (*PoP*). In this concept (Fig. 14.11), the expensive processor/ASIC is separately packaged in the PoP bottom package, while the memory is packaged in the
top package. Each of these two can be tested separately, while for the memory even
burn-in can be applied. Once both are proven to be fully functional, the parts can be
mounted on the application board. In the PoP concept, the stacked dies are connected through wires along their edges, which may increase the size of the package.
An alternative approach that is gaining increasing interest is the use of *through-
silicon vias* (*TSV*), which replaces the bond wiring by vertical connections through
the substrate of the dies.

TSV technology provides the possibility to connect completely different chips,
e.g. memories, microprocessors, RF and power management, in a much smaller
package than was possible with other 3-D techniques like MCM or PoP. Figure 14.12
shows the multilayer 8-gigabit DRAM TSV solution from Elpida as an example [54].

Future systems, however, will incorporate features and functional complexity
that will be even beyond today's imagination. They will combine the potentials of
physics, optics, mechanics, biology and chemistry with RF, analog and digital signal processing and storage capabilities packed onto one composite substrate. This is
usually referred to as *more than Moore* (Chap. 15). A target application may be a

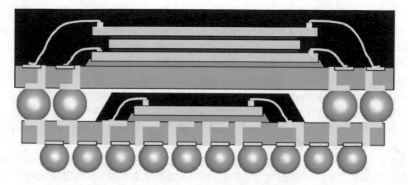

Fig. 14.11 PoP concept. (Source: www.emeraldinsight.com)

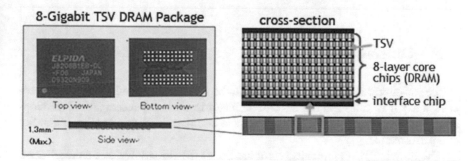

Fig. 14.12 Example of through-silicon via technology as a high-density packaging technology for DRAMs. (Source: Elpida)

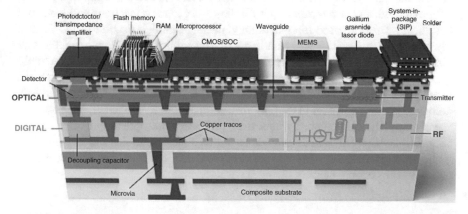

Fig. 14.13 Example of a system-on-a-package (SOP), consisting of optical circuits and devices, resistors, inductors, antennas, decoupling capacitors within a layered substrate and a combination of bare and packaged dies [55]. (graphic design by www.bryanchristiedesign.com)

device that could be encapsulated below the human skin to permanently monitor a person's health. This could be done by checking vital organs through the monitoring of breath, heartbeat, blood pressure, blood glucose level, etc. The results could be wirelessly communicated through the Internet to a medical advisor or physician, which can then propose the appropriate medical treatment, when necessary. In this case the sensors (monitors) may be attached on top of a substrate. Figure 14.13 shows an artist's impression of future *system-on-a-package* (*SoP*), which may combine optical circuits with passives, MEMs, SoCs and SiPs inside or attached to a composite substrate [55].

In conclusion, SoC and SiP are competing technologies, but this does not mean that there will be a winner in the end. SoC and SiP systems will live together but each in its specific application domains, where their properties are exploited to their full advantage.

14.8 Conclusion

While packaging was seen as a necessity to be tackled at the end of the design process, today it is more and more a critical factor towards the success of IC development. The package has an increasing impact on the overall chip quality, reliability and performance.

The growing IC complexity, combined with the drive towards miniaturisation and the continuous pressure on cost reduction, will not make the design process easier in the near future. In the coming years, it is expected that packaging technology will get closer to its limitations. For wire bonding, reductions in bond pad pitches will slow down or maybe even stagnate. Alternative bond pad layouts need to be explored, and new technologies for die-to-package connections will have to be developed. To follow the fab technology, package miniaturisation in the form of flip-chip, CSP and μBGA will gain in popularity, while on the other hand, the clear distinction between fab and assembly will fade. New vertical integration technologies have already been introduced, where substrate and chip technologies are used in combination with assembly techniques. SoC or SiP will stay as competing technologies, without a winner in the end. Instead, they will live together, but each in its specific application domains, where their properties are optimally exploited. One thing will be sure: SIP will be extended towards nonconventional technologies integrating MEMS-based applications, biosensors and/or optics. These devices are discussed in the next chapter.

Chapter 15
And, What Is Next?

15.1 Introduction

The continuous scaling of CMOS devices according to Moore's law has brought the design complexity in terms of number of transistors and performance requirements to an extremely high level. Design styles and methods need constant adaptation in order to manage this growing complexity and to enable full exploitation of the potentials of advanced CMOS technologies. A prediction of these potentials was presented in the International Technology Roadmap for Semiconductors (*ITRS*) [38], created by the Semiconductor Industrial Association (*SIA*). The final ITRS roadmap update was in 2013. Because of the changing semiconductor markets, from computer and consumer to mobile, automotive and IoT, the Semiconductor Industrial Association decided to change the focus of the ITRS towards seven topics and continued with an ITRS 2.0 version. With the end of Moore's Law at the horizon, ITRS 2016 has become the final roadmap and is replaced by a new initiative, named the *International Roadmap for Devices and Systems* (*IRDS*), which now also include *beyond CMOS* or *more than Moore* devices.

This chapter discusses the consequences of further scaling according to Moore's law (this is also called *more of Moore*), with a particular focus on the increasing complexities in all chip disciplines (lithography, fabrication, design, testing, packaging, etc.) which are approaching the fundamental physical limits.

The second part of this chapter is focused on what comes next. For many applications, 45 nm to 28 nm CMOS technologies will be sufficiently small to offer the required complexity, density, performance and power consumption. Particularly for the high-end microprocessor and memory applications, the scaling of CMOS technologies will still continue for a few more process generations. In the meantime, emerging micro-/nanoelectronic applications are being developed, which require new and disruptive device and integration concepts to create the systems of the future. Such applications include the integration of RF components, sensors (e.g. image, physical, bio, mechanical, electromechanical, chemical, temperature and

© Springer International Publishing AG, part of Springer Nature 2019
H. Veendrick, *Bits on Chips*, https://doi.org/10.1007/978-3-319-76096-4_15

pressure sensors) and actuators. The continuation of integration in this direction is generally referred to as 'more than Moore'.

While 'more of Moore' is more focused on further scaling, leading to smaller systems and improved computation efficiency, 'more than Moore' is focused towards improved interaction between user and environment.

15.2   Scaling Trends and Limitations: *More of Moore*

15.2.1   *Introduction*

Nanometre CMOS IC design requires more focus on the physical design and on the consequences of further scaling. This will certainly have an impact on the semiconductor technology roadmap. In the race towards a multi-giga-transistor heterogeneous system-on-a-chip (SoC) (see Fig. 15.1), design methods and tools not only have to be changed to make the design manageable (system design aspects) but also to make a functional design (physical design aspects).

The complexity of such a SoC can only be managed by applying:

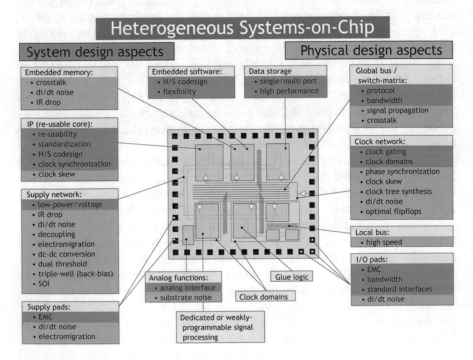

Fig. 15.1 Important aspects of a (heterogeneous) system-on-a-chip

- A platform with integrated hardware/software architectures and application development tools
- Increased design automation: system-level synthesis to improve design efficiency
- Reuse of predefined logic, processor, memory and interface cores
- More design resources per chip

The first three items deal with system-level design aspects. The increased design resources, however, are not only required to manage the SoC design complexity; they are also needed to cope with the increasing number of physical design aspects as depicted in the green boxes in the figure. Previously, only analog circuits were susceptible to these physical effects. In future process generations, these effects will dominate the SoC's performance and signal integrity, while some of these effects are already threatening the performance of complex VLSI chips for more than a decade. Future chip design therefore requires a more analog approach. Design is no longer about switches and ones and zeros only, but also about resistors, capacitors, inductors, noise, interference and radiation.

Basically, a VLSI chip is just a bunch of transistors that perform a certain function by the way that they are interconnected. The next sections show how the increasing complexity impacts the various semiconductor disciplines.

Particularly the increasing design complexity has caused a paradigm shift in the cost components of an IC: from fabrication costs towards design costs.

15.2.2   System Complexity

State-of-the-art chip fabrication technologies enable the integration of a *system-on-a-chip (SoC)*, with hundreds of millions to several billion transistors. This has led to the convergence of consumer, computing and communication domains, which accelerates the introduction of new features on a single chip. As a result, systems and chips become more heterogeneous (Fig. 15.1), i.e. they will consist of an increasing variety of application specific and general-purpose IP cores, memory cores and interfaces. On its turn, this leads to different performance demands for different parts of the system or even for different parts of a single chip. The system must offer the best performance and the lowest possible power, which are conflicting demands that can only be met by solving the related problems at all levels of design, starting at system level. System designers must therefore exactly know the impact of their choices on performance, power and size of the system, such that they can ease the tasks of the IC, board and package designers.

Finally, the large variety of functions, running at different performance and power levels, requires very extensive and time-consuming hybrid simulations and emulation to still guarantee the system to operate correctly under all operating conditions. System validation and verification have become serious parts of the overall system design and implementation.

15.2.3   Design Complexity

It is not only the considerable amount of transistors that makes chip design an increasingly complex and time-consuming task. The demand for more features at increased performance has led to unacceptably high-power consumption. This has forced all disciplines that contribute to the overall system development, to intensify the search for power reduction techniques both in process development and at all levels of design.

The combination of high-performance and low-power circuits to support different functions on a single chip has resulted in ICs with different voltage and/or frequency domains. Chips with 30 or more different clock domains are no exception. However, communication between these different domains must still be guaranteed. A system-on-a-chip also requires the integration of digital, memory, analog and RF parts. The simultaneous switching of large amounts of digital circuits is causing increased noise peaks, while the analog circuits are very susceptible to them. Next to this, the physical spacing between interconnect lines is also reducing, which may lead to increased crosstalk between signals over these wires. In general, due to the increasing noise levels, both digital and analog designers have to invest more design time in achieving noise reduction in the digital parts and reduced noise sensitivity in the analog parts (Fig. 15.2).

While digital circuits are normally designed using minimum layout design rules (minimum allowed physical sizes), most analog circuits on an SoC use larger-than-minimum design rules to make them less sensitive to process variations and more robust to noise. The huge number of transistors, the increasing number of voltage and frequency domains and the increasing diversity of mixed-signal circuits causes the verification and test development to be also a major part of the total design effort and overall cost of an IC.

Fig. 15.2 Symbolic representation of a mixed analog/digital IC

15.2.4 Silicon Complexity

Current semiconductor fabrication processes enable the production of processor ICs with several billion transistors, while advanced flash memories even contain close to 100 billion transistors. The continuous scaling of the transistors in conventional silicon technologies has come to an end. These technologies used MOS transistors with silicon-dioxide as gate dielectric and with aluminium as interconnect. For many process generations in the past, every new technology node yielded an intrinsic circuit speed increase between 40% and 80%. However, CMOS circuits beyond the 180 nm node show increased manifestation of physical effects, such as short-channel effects, increased subthreshold leakage currents and carrier mobility reduction, which reduce the transistor current. Therefore, speed increase between successive process generations, since then, was no longer a given fact. This is one of the reasons why the speed (clock frequency) of microprocessors has almost saturated to between 3.5 and 4 GHz and the emphasis in performance competition has moved away from gigahertzes to multicore architectures [56] (see also Part I, Chap. 1).

In many processes beyond 28 nm, silicon-dioxide as gate dielectric has been replaced by a *high-k dielectric*, while the aluminium interconnections have been already been replaced by copper interconnections since the 120 nm node. These are isolated by a *low-k dielectric* material to reduce mutual capacitance for power and noise reduction and performance increase. Additional process steps, such as built-in transistor channel stress, additional local doping, etc., are needed to further enhance performance. Transistor currents depend on the amount of doping atoms in the transistor channel. This dope fluctuates from transistor to transistor on the die, from die to die, from wafer to wafer and from batch to batch. Next to the doping statistics, also the physical sizes of the transistors and interconnections are not everywhere perfectly reproduced. These more or less random *process variations* are generally referred to as *variability* or *process spread*. Variations in transistor currents depend on these *random doping fluctuations (RDF)* and statistics. Because transistor sizes decrease, the number of doping atoms in a transistor channel decreases, while the variations in transistor currents increase, and performance predictions are much more difficult to make.

Continuous scaling of the physical sizes also affects the reliability of a chip. An example is the scaling of the widths of metal interconnections. This is quite the opposite from what is required by the increasing chip currents. Increasing currents through narrower metal lines lead to increasing current densities in these lines. A current is just a flow of electrons. Above a certain current density (read, electron density), the metal ions in the metal line tend to move in the direction of the electrons. Physically these metal ions then migrate from one position to another. This effect is called *electromigration*, which may cause a huge reliability problem if it is not yet taken care of in the design by verifying the current densities in the supply lines to the various on-chip cores.

15.2.5   Fabrication Complexity

The early metal oxide semiconductor (MOS) processes required only four to five masks to fabricate a chip. Particularly the increasing number of metal layers, from 1 in the 1960s and 1970s to 8 or 12 today, contributed to the large increase in the number of mask layers. Each additional metal layer requires two more masks: one for the metal layer itself and one for the contacts (vias) to the previously processed metal layer. Next to that, the growing diversity of on-chip devices (analog, digital, memory, RF and interfaces) and additional process steps to improve performance and reliability also require several additional masks. The production of modern SoC ICs, therefore, requires between 30 and 40 masks, or even more in case multi-patterning lithography techniques are used. In the chapter on lithography, it has already been explained that there are relatively cheap binary masks for those patterns that do not need the minimum process feature sizes. However, the masks that define the physical sizes of the transistors and interconnections require an extremely high level of accuracy and are therefore very expensive. With the reducing feature sizes, also the photolithography process and tools to copy the mask image to the wafer become dramatically more complex and expensive. In sub-28 nm processes, in which double patterning is used, the number of masks may even exceed 60.

Moreover, smaller feature sizes on the chip and on the mask increase the impact of dust particles during the photolithographic process, as these are also imaged onto the chip, causing shorts and/or opens, which may reduce yield or lead to reliability problems.

All in all, the continuous miniaturisation requires:

- Increased accuracy during the production of the transistors and interconnections leading to a better control of almost all process steps
- Improved resolution of the lithographic systems
- Higher level of cleanliness of the clean rooms
- Increased process monitoring during the production process
- Increased skills and higher level of discipline of the process operators

These requirements have a severe impact on the total costs of a fully equipped wafer fab. A discussion on the costs is presented further in this section.

15.2.6   Package Complexity

The large number of devices (processors, memories, interfaces, RF, power management, etc.) that a chip has to communicate with, combined with increasing bus widths, requires that the ICs and their packages have to deal with an increasing amount of power and I/O pads. Even some consumer and telecom chips can have between 1000 and 2000 pins.

Electronic gadgets, such as tablet computers and smartphones, integrate an increasing diversity of functions but still require very small physical sizes. This is a drive to move from lateral package solutions, such as multi-chip modules (MCMs), to 3-D packing and stacking technologies, such as system-in-a-package (SiP), package-on-a-package (PoP), system-on-a-package (SoP) and through-silicon vias (TSV) (see Chap. 14). These relatively new technologies must still match the stringent and increasing electrical, thermal and reliability requirements. Over the last couple of decades, there has been only limited improvement in thermal conductivity of the packages. Also stacking of packages increases the complexity of the thermal management at system level. It requires 3-D thermal models of the packages and the in-between dielectrics.

15.2.7   Test, Debug and Failure Analysis Complexity

After the chip has been processed, its functionality must be verified and tested. When a chip fails, particularly at first silicon of a design, extensive debug and failure analysis techniques are needed to find the location and cause of the failure. Because the scaling factor between two successive IC technology nodes is close to 0.7, every new IC technology node enables the integration of two times more transistors on the same chip area. However, most ICs have their terminals (pads) still at the periphery of the chip, such that the number of pads increases by only 40%. So, every new technology node, about two times more transistors have to be observed (tested) through only 40% more pads, which makes it much more difficult to find the relation between a chip's malfunction and the failing circuit that causes it. Also the increasing number of interconnection layers on a chip prevents easy access to the actual circuits. Very advanced failure analysis tools are therefore required to enable observability from the backside of the chip. Use of these tools is costly and time-consuming. Design for test and design for failure analysis are more or less common practice in the development of SoCs to improve observability and reduce time to market and costs. Test and failure analysis have extensively been discussed in Chaps. 12 and 13, respectively.

15.2.8   Time-to-Market

Simultaneous with the explosion in chip complexity is the required reduction in *time-to-market* (*TTM*) for many application areas, particularly in the rapidly changing communication and consumer markets. The commercial success of a design project depends to a large extend on the design team's skills and ability to deliver their chip on time. This requires well-disciplined and highly skilled IC designers, with an extensive knowledge of the potential pitfalls in first silicon. Already today, between 40% and 50% of the total IC development time is spent on validation and verification tasks. Even then, roughly two third of all IC design projects exceed their original time schedule. Without adequate top-level-verification methodologies and sufficient test coverage, errors can easily go undetected. Due to the increase in design complexity, more than 60% of first silicon (year 2005) did not work properly. It is assumed that that number has not reduced during the last decade, since the design complexity has exploded. Such a *respin* may roughly take a couple of months to half a year, which could lead to a dramatic loss in market share. Most of the respins are due to logical/functional errors, followed by timing/performance failures and analog performance.

'If you ask people if their chips work first time, you will have a hard time finding someone who will admit to a respin or two, although we expect multiple silicon passes are more common than is advertised' [57].

This continues the need for complete validation and verification methodologies and tools. But, even then, the next designs in the next technology node would be even more complex and would still require more advanced verification methodologies and tools. In other words, as long as Moore's law remains valid, design projects will still miss their original schedule and might still require one or more respins, despite the time-to-market pressure. The ITRS states, 'Without major breakthroughs, verification will be a non-scalable, show-stopping barrier to further progress in the semiconductor industry'.

15.2.9   Cost

The design takes an increasing portion of the overall chip development cost. This is due to the dramatic explosion in design complexity, as discussed before, and in the number of design tasks. Design for test, design for robustness, design for debug, litho-friendly design, design for manufacturability, etc. all add up to an increased design complexity and chip area. The level to which these additional measures will limit the efficient use of chip area cannot be predicted because it also depends on the creative design alternatives that will be developed in the near future. Figure 15.3 shows a summary of increasing design tasks. It also shows the exponential increase in average ASIC design cost (original Source: IBS2009; diagram has been extended based on many semiconductor publications at conference and in magazines) because of the rapidly increasing design complexity.

These costs include the complete system, the architecture and software development and reflect the average ASIC category [2]. For a 32 nm ASIC, these may rise around 75 million US$ [58]. Assuming that the chip is meant for a consumer application and that the earnings per device are expected to be in the order of 1US$, then a simple calculation learns that at least a total volume of 75 million devices is required to reach breakeven with respect to the development cost. There are not so many applications that generate market volumes for a single supplier in this order of magnitude. For the sub-20 nm nodes, the total development cost only increases further.

Figure 15.4 shows that the total design cost is increasing much faster than the other cost contributors. A set of about 45 masks for a mature 65 nm CMOS

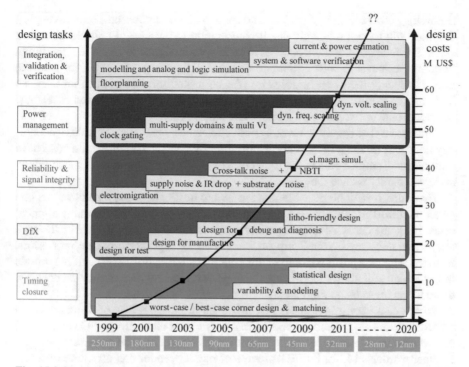

Fig. 15.3 Increasing number of design tasks and growing design costs. (Original Source: IBS 2009, but diagram has been extended based on many semiconductor publications at conference and in magazines)

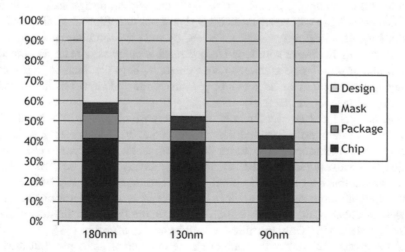

Fig. 15.4 Various contributions to the overall chip development costs for different technology nodes. (Source: Leon Stok (IBM) ISPD2003)

technology cost around $0.5 million and about $1.5 million for the 28 nm node [59]. But they still have a relatively small contribution to the overall chip development costs. The total development costs of a 1 cm² chip, designed to run at the maximum frequency achievable with a standard-cell design flow (e.g. 700 MHz for a 45 nm SoC), including the costs of software (for both platform and application), are so high that the required volume to recover the development and fabrication costs must be about equal to the number of transistors on that chip.

This means that soon, for several application areas, the move to the next technology node may no longer be economically attractive. When the fabrication costs have only become a minor part of the total costs of an IC, scaling to the next technology node will hardly reduce the price.

Therefore, it may be possible that the 28 nm or 22 nm node, plus or minus one generation, will become the last economically viable technology for many applications. For some cheap consumer products, the 45 nm node may already be the final one, while this may be the 7 nm or 5 nm node for high-performance processors. Both the price and profit per chip in this category are roughly an order of magnitude larger, which still allows recovering the huge development and production costs for these ICs in these technology nodes.

What's also very important is that below 20 nm there is less of standardisation regarding feature sizes and process nodes. This means that different IC vendors can have different process node names for processes that contain the same feature sizes. In other words, a 14 nm FinFET process of one IC vendor can use quite different minimum feature sizes than a 14 nm FinFET process of another vendor. This makes further area and performance comparisons complicated. The 7 nm and 5 nm process node names, mentioned above, refer to real channel lengths. What is called a 7 nm process, today, may have channel lengths of 10 nm or more.

The ability to completely verify, test, debug and diagnose future complex designs will reduce dramatically. It is therefore likely that current design styles with fixed and dedicated logic will be replaced by design styles that allow flexibility and configurability. This flexibility can be enhanced by software solutions (programmability) as well as hardware solutions (*reconfigurable computing* such as embedded FPGA and/or sea-of-gate architectures). Remaining bugs can then be bypassed by changing the program or by remapping (part of the function) on the embedded reconfigurable logic, respectively.

Another potential key factor in lowering the pace of scaling is formed by the economics of the production facilities. From 1966 to 2017, the costs of a chip production plant increased by a factor of 3000, from about $5 million to close to $15 billion, respectively (see Fig. 10.14). There are already several new wafer fabs in development that cost over $20 billion. These investments can only be raised by a few individual large semiconductor companies and semiconductor alliances. The process of outsourcing chip fabrication and becoming fab-lite or fab-less will continue in this decade. 'Only the elite few will be able to afford it' [58].

An analysis [72] shows that, although the number of gates per unit area still increases with further scaling, the cost per gate saturates around the 28 nm node (Fig. 15.5). It is also expected that the cost per gate for the 16/14 nm FinFET nodes is higher than for 20 nm and 28 nm.

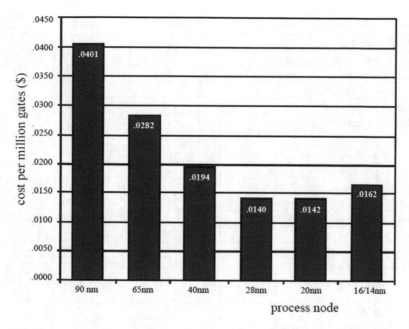

Fig. 15.5 Expected trend in reduction of cost per million gates, which seem to saturate at the 28 nm node [72]

It has already been discussed before that, when the design costs become significantly larger than the fabrication costs, the drive for scaling an application to the next technology node will reduce. Moreover, since electrons run close to their saturation speed in the 65 nm technology node and beyond, only limited circuit performance improvements can be expected from this scaling. This has changed the focus from GHz to multicore designs.

All in all, the semiconductor design and manufacturing landscapes are changing quickly, because almost all semiconductor disciplines are approaching their complexity limit.

15.2.10 Conclusion

Moore's law (a quadrupling of IC complexity every 18 months to 2 years) has proven its validity for almost six decades, now. In this period, this complexity has grown from a few transistors to several billions. It is sometimes called a self-fulfilling prophecy. It is viewed as a measure for future trends and sets the pace of innovation. Almost according to this law, the Semiconductor Industrial Association has set up its roadmap for the next couple of years. Table 15.1 shows several important parameters of this ITRS roadmap. It is based on data from [38] and [60].

This section has shown that the semiconductor industry has reached the complexity limits of nearly all of its contributing disciplines. An example of reaching

Table 15.1 Important IC characteristics and their change based on [38] and [60]

Year of production	2013	2015	2017	2019	2021	2023
Logic industry node range [nm]	"16/14"	"11/10"	"8/7"	"6/5"	"4/3"	"3/2.5"
Minimum feature	29	23	18	14	11	9
Low-power (LP) roadmap						
Physical gate length (LP) [nm]	23	19	16	13.3	11.1	9.3
Effective channel length (LP) [nm]	18.4	15.2	12.8	10.6	8.9	7.4
Supply voltage [V]	0.86	0.83	0.8	0.77	0.74	0.71
Equivalent oxide thickness (EOT) [nm]	0.8	0.73	0.67	0.61	0.56	0.51
Physical gate oxide thickness [nm]	2.56	2.53	2.49	2.42	2.37	2.29
Dielectric constant of gate dielectrics	12.5	13.5	14.5	15.5	16.5	17.5
Channel doping bulk/metal gate [10^{18}/cm^3]	5/0.1	7/0.1	8.4/0.1	?/0.1	?/0.1	?/0.1
Mobility bulk/metal gate [cm^2/Vs]	400/375	400/375	400/375	?/375	?/375	?/375
Metal gate I_{off} [pA/µm]	10	10	10	10	10	10
Metal gate drive current $I_{ds,sat}$ [µA/µm]	643	618	574	550	537	458
MG nMOSt intrinsic delay ($\tau = CV/I$) [ps]	1.62	1.53	1.59	1.53	1.42	1.47
nMOSt dynamic power indicator (CV^2) [fJ/µm]	0.9	0.81	0.73	0.65	0.57	0.48
Maximum chip power [W]	3	3	3	3	3	3
High-performance (HP) roadmap						
Physical gate length (HP) [nm]	20	16.7	13.9	11.6	9.7	8
Effective channel length (HP) [nm]	16	13.4	11.1	9.3	7.8	6.4
Supply voltage [V]	0.85	0.8	0.75	0.7	0.66	0.62
Equivalent oxide thickness (EOT) [nm]	0.8	0.73	0.67	0.61	0.56	0.51
Physical gate oxide thickness [nm]	2.56	2.53	2.49	2.42	2.37	2.29
Dielectric constant of gate dielectrics	12.5	13.5	14.5	15.5	16.5	17.5
Channel doping bulk/metal gate [10^{18}/cm^3]	6/0.1	7.7/0.1	9/0.1	?/0.1	?/0.1	?/0.1
Mobility [cm^2/Vs]	400	400	400	?	?	?
Metal gate I_{off} [nA/µm]	100	100	100	100	100	100
Metal gate drive current $I_{ds,sat}$ [µA/µm]	1670	1700	1660	1600	1450	1330
MG nMOSt intrinsic delay ($\tau = CV/I$) [ps]	0.569	0.525	0.469	0.474	0.477	0.437
nMOSt dynamic power indicator (CV^2) [fJ/µm]	0.82	0.74	0.66	0.58	0.51	0.41
Maximum chip power [W]	156	144	130	130	130	130
MPU/ASIC Metal 1 (Ml) ½pitch [nm]	40	32	25.3	20	15.9	12.6
Chip size [cm^2]	5.8	4.1	4.1	4.1	4.1	4.1
Number of transistors [billion]	14.7	24	35.4	60	100	141.5
Transistors/cm^2	3.5	5.5	8.6	14.7	22.5	35.3
Number of bondpads	5200	5600	6000	6500	6700	6840

the physical limits of scaling is the flash memory. Around the 1990s, each cell contained about 1 million electrons to represent a logic '1'. Today, this has gone down to only a few hundred. This cannot continue forever [61]. The limitation of scaling planar flash memories is also reflected by using the third dimension by creating the memory cells in many stacked layers (close to 100 layers in the year 2017!!) above each other on the same chip.

Particularly the complexity of the design, verification and test tasks has accelerated over the last decade and forms a potential barrier to obtaining full exploitation of the available manufacture potentials. Moreover, they also contribute to the exploding cost of IC development. The overall success of the semiconductor industry will be increasingly dominated by how the difficult design, engineering and test challenges will be addressed [38].

Although the physical limits of transistor operation lie close to 7 nm (real channel length), for most applications the limits of scaling will be reached earlier. Today, we already face the fact that a move to the next process generation is no longer commercially attractive for various application domains. For cheap high-volume consumer products, however, this point in time will already be reached within a couple of years.

So, what's next?

15.3 The Next Decade: *More than Moore*

15.3.1 Introduction

Scaling cannot go on forever. At some point in time, sizes will reach molecular and atomic levels. From the previous sections, it is clear that already during the last decade, the semiconductor industry was fighting a harder battle against the exceptionally high complexity in all of their disciplines. This trend is expected to continue with Moore's law (more of Moore) slowing down and approaching its final stage, most likely in the next decade. In the meantime, a paradigm shift is taking place in semiconductor technology. The slowing down of Moore's law is compensated by an increasing combination of silicon-based micro- and nanoelectronic technologies, micro- and nanoelectromechanical systems (*MEMS* and *NEMS)*, microchemical systems (*MiCS*) or microbiological systems (*MiBS*), sensors and/or actuators.

This enables the integration of sensors (feel vibration, pressure or temperature; taste liquids; smell gases; see images; hear sounds) or actuators (act upon a sensed signal) and electronics (who are the brains of the system) on the same substrate. In many publications, MEMS cover also the microchemical and microbiological systems, and as such, we will follow this approach. In other words, whenever relevant, the term MEMS here also includes the MiCS and MiBS.

While microelectronics integrates electronic devices on a common substrate, fabricated with IC technologies, MEMS integrate a number of microelectromechanical devices on a common substrate using compatible microfabrication technologies. Combining both technologies enables the integration of true systems on a single chip. The sizes of most of today's MEMS are usually in the range of several microns to a few millimetres. However, a lot of R&D is currently devoted to *nano-technology*, which is a buzzword for a large range of technologies that enable device sizes in the order of nanometres. Some of these technologies can manipulate materials and totally change their properties, while others will further reduce the size, power and costs of future materials, devices and systems. The scope of nanotechnology is much broader than that of microelectronics and MEMS, but it will enable MEMS with nanometre sizes. These *nanoelectromechanical systems* (NEMS) are seen as the logical successor of MEMS. The impact of nanoelectronics and nano-technology on society, environment and economy will be even more than micro-electronics has done so far.

The following subsections present a summary of sensors, actuators, MEMS and nanotechnology together with a grasp of their current and future microelectronic-related applications, in order to create a picture of what lies ahead of us.

15.3.2   Sensors, Actuators and MEMS

Comparable to microchips, also sensors and actuators are everywhere, even when we are unaware of their presence. Both sensors and actuators enhance the quality of our everyday life by improving comfort and safety levels. It is neither the intension of this chapter to explain the operating principles nor application areas of all of these devices. It is much more important to show the reader the large variety of these devices that are currently already in use, in order to create a vivid picture of the enormous potential of these devices when scaling the technology into molecular or atomic sizes.

A *sensor* converts physical inputs (stimuli) into electrical signals that can be further processed by electronic circuits. A well-known sensor that is already in use for a couple of decades is the image sensor. Figure 15.6 shows a photograph of an 24 million-pixel CMOS image sensor. The light is captured by the pixels in the sensor and converted into small electrical charges. In fact, such a sensor converts an optical image into an electrical image. When the image is read, the pixel charges are converted into voltages and then digitised and stored in a memory.

The penetration of sensors in our lives happens so fast that we are not aware of many of their applications. Table 15.2 shows a summary of the most frequently used sensors in a wide range of applications.

Image sensors are also inserted under the retina of people who suffer from reti-nitis pigmentosa, which is an inherited disorder which gradually destroys the retina [62].

An application area where sensors are already widely utilised is the automotive industry. As the demand for car safety and convenience grows, so does the need for reliable and accurate sensors. By the year 2020, it is expected that a typical car may contain more than 200 sensors and offers numerous potential opportunities for further sensor applications. Figure 15.7 shows the penetration of sensors into the car [63].

Fig. 15.6 Example of a 24 million-pixel image sensor (≈ 9 cm^2) for digital photography in LCC package for Leica Camera. (Courtesy of CMOSIS)

Table 15.2 Examples of most frequently used sensors today

Sensor	Application
Magnetic field	Compass (also in a smart phone)
Temperature	Thermostat (home, car, weather, pc, water boiler, etc.)
Light	Night light, smoke detector
Infrared light	Motion detector, IR-camera, security systems
Electro-optical	Night light, emergency lamps
Ultrasound	Motion detection, burglar alarm, security systems, car distance control
Radar	Motion detection, speed control, automatic door openers
Image	(Film)camera, security systems, car parking sensors, face detection
Pressure	Weather, airbags, altitude (aviation), tire pressure, blood pressure
Vibration	Process control, burglar alarm, shock detector, accelerometer, microphone, airbags, cell phones, camera's (image stabilisation)
Humidity	Weather, environmental control, rain sensor
Flow	Gas meter, water meter, etc.
Electrical current	Electricity meter, battery chargers, over-current protection

A sensor in itself does not improve the driving experience. Every sensed signal must be processed or displayed in order to support the convenience, comfort or safety of the driver. The electronics in a typical car, today, contains hundreds of chips to process all the data coming from the sensors. In fact, a car is a computer on wheels: the ultimate mobile computer.

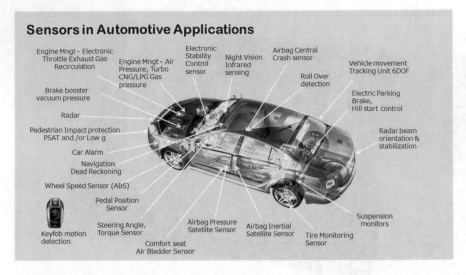

Fig. 15.7   The sensor penetration into a car. (Source: NXP Semiconductors [63])

There is currently a lot of talking on *autonomous cars* (*self-driving cars*), which currently contain more than 100k US$ of sensors and electronics. In all circumstances, the electronics must continuously translate the sensor data into extremely accurate perceptions, predictions and 100% correct decisions. In fact, such a car is constantly self-learning and improving based on complex machine learning algorithms. This is a form of *artificial intelligence* (*AI*), which may lead to misperceptions. Every human being is a self-learning person. However, we don't put anyone behind a steering wheel of a car before the age of 16 and after we have tested her/his accumulated driving knowledge and experience. This is quite different from an autonomous car, which must take a 100% correct decision right from the first minute. Therefore, not all publications on autonomous cars share the same optimism [64]. We may expect the introduction of fully autonomous cars on only a limited set of predefined routes as taxis or buses. It may therefore take many more years before they become available for consumers [65].

Another example of the potential use of sensors is in so-called *smart roads*, which may include sensors to monitor road conditions and traffic patterns. These roads of the future can also be combined with *IoT* technology, using solar-friendly material to power various electronic devices and features to monitor traffic speed, weather conditions and vehicle-to-road and vehicle-to-vehicle connectivity in order to improve the safety and efficiency of the roads [66].

The total shipment of automotive sensors in 2018 is close to 25 billion US$.

As said before, a sensor transfers physical inputs into electrical signals. On the contrary, an *actuator* is a device that provides physical responses to (mostly) electrical inputs.

The list of actuators (Table 15.3) is somewhat shorter than that of sensors but still shows a relatively broad range of applications. Only those actuators that have a certain relation with electrical signals are included in the table.

Table 15.3 Examples of most frequently used actuators today

Actuator	Application
Stepper motor	Positioning systems, pick and place equipment, printers and scanners
Thermal switch	Temperature control, thermostat, safety shut-off devices
Light	LED, display
Electromagnetic	Doorbell, speaker, fuel injectors, dot matrix printers
Electrostatic	Resonators, oscillators, gyroscopes, micro-mirrors, x-y micro-stages, speaker
Piezoelectric	Micro-manipulators, vibration generators, image stabilisators
Magnetostatic	Relay switches, barcode readers, biomedical devices
Hydraulic	Brake, draglines, clam-shell bucket
Pneumatic	Hammer, air compressor

Many of these sensors and actuators are realised with conventional and bulky technologies, but due to the increasing progress in MEMS and nanotechnology, many of these devices can be scaled to milli-/micro-/nanometre sizes. There are two well-known examples of sensor/actuator devices. The first one is a closed-loop video system where a camera (sensor; Fig. 15.6) captures the image and transfers it into electrical signals, while a display (actuator) converts them back into the image. The second example is a hearing aid. It contains a microphone (sensor) to pick up the sound signal and converts it into electrical signals, a processor that does some special audio processing, which can adapt to the listening environment, an amplifier to intensify the signal and an extremely small loudspeaker (actuator). Also here the progress in both chip and MEMS technology has led to huge improvements in the quality, size, power consumption and usage of hearing aid devices, which are placed behind or in the ear. For persons who are profoundly deaf or experience a severe loss of hearing and have had little or no improvement using hearing aids, cochlear implants may gain or restore (part of) the hearing ability. Figure 15.8 shows the external and implanted components of a cochlear implant.

Most of the hearing loss or deafness is caused by missing or damaged hair cells in the cochlear. These hair cells are the sensors to stimulate the auditory nerve system. A *cochlear implant* is an electronic device that bypasses a large part of the 'ear system', including the eardrum, middle ear and these damaged sensors. It directly stimulates the auditory nerve system, which transfers the signals to the brain, where they are perceived as sound. Although this sound is different from normal hearing and takes time to learn, it enables hearing and understanding speech and allows participation in conversations.

As said before, current MEMS may have dimensions in the order of microns. Original MEMS technology was very similar to IC process technology. They were often manufactured on silicon substrates using well-established IC processing steps. Next to the commonly used materials in IC fabrication, MEMS may also use other materials, such as ceramics and polymers and, more recently, carbon nanotube arrays as well. Figure 15.9 shows an example of a MEMS accelerometer. During acceleration, the mass deforms the springs, such that the capacitances between the fixed and the moving fingers change proportionally to the deformation of the

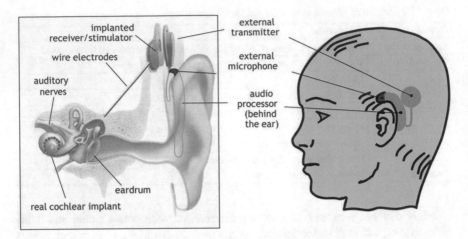

Fig. 15.8 The components of a cochlear implant and their physical position. (Source of ear cross section: https://www.healthbase.com/hb/pages/EarNoseThroat_ENT.jsp)

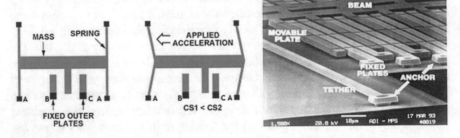

Fig. 15.9 Structure of a MEMS accelerometer. (Source: Analog Dialogue 43-02, February (2009))

springs. The electronic circuits, which are connected to the terminals A, B and C, detect these changes and process the results.

MEMS can be implemented with the electronic and mechanical parts integrated on the same substrate as a single-chip solution or as a multi-chip solution with separated MEMS and ASIC dies. In case of a single-chip solution, the mechanical parts can be positioned next to or on top of the electronic parts, depending on the type of sensor or actuator. Single-chip solutions are only used in applications in which they are cost-effective and/or where they fulfil the requirements on area (volume) minimisation. The complexity of MEMS is also a big concern. MEMS design is quite different from IC design. First, a MEMS designer must be familiar with the various disciplines involved in the MEMS operation and applications, as discussed in Sect. 15.3.1.

Many traditional IC design houses have become fab-lite or fab-less. The lack of access to a production facility could act as barrier for them to start a new development in MEMS, while it may enable newcomers to penetrate the market.

The large variety in MEMS technologies and their ability to sense (feel, see, hear, taste, smell) or actuate (press, move, rotate, switch, display) a particular parameter put stringent demands on the packaging technologies, as they have a lot of impact on the performance and eventual physical size of the MEMS system. Moreover, MEMS are often used in hostile environments where they are subjected to stress, humidity, interference, deformation, vibration and/or temperature cycles, while they must perform accurate and reliable in all circumstances. This is quite an engineering challenge, which requires the availability of sophisticated analysis tools with accurate models of the involved physical mechanisms and applied packages.

15.3.3   Nanotechnology and NEMS

Although some publications interchange MEMS with nanotechnology, in this book nanotechnology is seen as a combination of a large variety of nanoscale technologies. Nanotechnology is currently expected to be the most potential successor and/or compliment technology of the current semiconductor technology. Traditionally, *nanotechnology* is defined as a technology targeting at dimensions between 0.1 and 100 nm (1 nm is one billionth of a metre; the thickness of a human hair is about 100,000 nm; see Fig. 1.14 in Part I) that enables the development of materials, structures, devices, interconnections and systems with new properties and functions. It will lead to the development of fundamentally new products and applications. Today nanotechnology is much more identified with atoms and molecules. Its impact is economical, industrial, environmental as well as societal.

Billions of dollars of government money are yearly spent on nanotechnology research. This has resulted in the fact that nanotechnology is already applied in a broad scope of products: paint, batteries, lighting, integrated circuits, tennis balls and rackets, fishing rods, snow boards, skis, tires, water purification, sunscreens, disinfectants, medical implants, food, etc. Thousands of patent applications are filed every year. The technology is expected to make existing things smaller, faster, lighter or stronger or to create totally new products. It will affect many aspects of our lives: labour, health, clothes, food, communication, housing, mobility, energy, etc. These examples show that the new applications are the result of intensive interdisciplinary cooperation of physicists, chemists, engineers, biologists, biotechnologists, medical, biomedical and material scientists. The wide scope of expected new applications of nanotechnology includes short-term and long-term applications.

Certainly not all of its applications (will) have a relation with (micro-/nano-) electronics. Of those, who have, we distinguish nanotechnology-enabled MEMS (including MiCS and MiBS) sensors and/or actuators, bio- and carbon nanotube transistors and wires and single-electron devices. Sometimes the nanotechnology-enabled MEMS are called *nanoelectromechanical systems* (*NEMS*). Many of these potential nanotechnology applications have a long-term scope. Others show a higher degree of realism and have short-term potentials. All in all, the fast progress in MEMS and nanotechnology will create unlimited opportunities in the world of sensors and actuators. It is expected that these devices will find their way into all commercial, medical and military markets in the near future. In many of the existing applications, such as automotive and communication, these nano-systems will create new functions and features to increase comfort and safety and/or increase the level of integration to further reduce the size, weight, power and cost.

However, a relatively new application area that is expected to particularly benefit from this progress is biology with an emphasis on biotechnological and biomedical applications. Biotechnological applications include diagnostics, DNA identification, drug detection, food inspection, bacteria and virus detection, etc. Biomedical applications include biotelemetry (to remotely monitor patients), drug delivery, precision surgery (use of robots), etc.

Just to get an impression of what lies ahead of us, we will present a few examples of the progress in the biomedical arena.

The combination of nanotechnology-enabled biosensors/actuators and integrated circuits leads to the development of miniature biomedical analysis systems with very high potentials. Several technological approaches are expected to gain significant market share by replacing expensive and bulky conventional lab equipment by so-called biochips. *Biochip* often refers to a broad scope of technologies. They consist of a biological entity and a semiconductor chip. Basically, these biochips contain a miniaturised lab that can perform many biochemical analyses at the same time. An example is the *lab-on-a chip* (*LoC*) technology. An LoC is a single-chip device of only a few square millimetres or centimetres that can perform one or more laboratory functions on extremely small volumes of fluids, often only several nanolitres. It can be used for diagnostics in various clinical and life science applications. Today's LoC devices are of limited complexity and are relatively small. Figure 15.10 shows a few recent LoC examples. It is beyond the scope of this book to explain their functionality in more detail. The interested reader can find much more information on the Internet.

Another application area of biochips is in the biotelemetry. An example of this is the use of an endoscope (Fig. 15.11 (left)) in gastroscopy: the inspection of the gastrointestinal tract.

The conventional endoscope consists of a long and flexible tube with a built-in camera and light at one end. The camera captures images, which are displayed by a

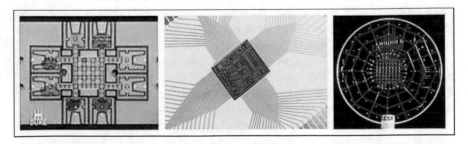

Fig. 15.10 A few examples of existing lab-on-a-chip devices: (**a**) device for monitoring enzymes in the bloodstream or measuring harmful chemicals in the environment [67], (**b**) clinical diagnosis chip [68], (**c**) two-dimensional micro-valve array technology [69]

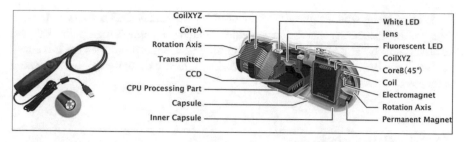

Fig. 15.11 Conventional endoscope (left) and next-generation capsule endoscope (right; Source: RFsystemlab [70])

computer monitor. The tube contains various channels. The light is often transferred through an optical fibre. An additional channel allows entering medical instruments and manipulators to enable a variety of medical services, such as biopsy and removal of foreign tissue, such as tumours. Today, there is a lot of research towards the integration of medical diagnostic devices. The *lab-on-a-pill* (*LoP*) approach [70, 71], also called a diagnostic pill (Fig. 15.11 (right)), may replace the conventional endoscope, particularly for diagnostic purposes. This pill is designed to be swallowed and will perform measurements when passing the gastrointestinal system. It contains a couple of LEDs to illuminate tissue walls, a CCD camera chip, a CPU that performs signal processing and a transmitter for the wireless communication (telemetry). It has no battery, as the patient wears a vest, which contains a coil that generates an electromagnetic field. The pill consists of two capsules, an outer and an inner one. As the outer capsule travels through the gastrointestinal system, the external induced electromagnetic power turns the inner capsule, including the CCD camera, around. This camera is facing the side and spinning 360 degrees so that it scans through the locally transparent capsules, directly over the tissue walls. It completes a full swing every 12 s for repeated close-ups. It takes about 2 min to travel 1 inch [70] and about 8 h before leaving the body.

Most of these advanced LoC and LoP developments are still in the research or prototyping stage, but with an ageing population and a strong demand for cost reduction in public healthcare, it is expected that these biochips will soon find their way into home monitoring and diagnostic applications.

As indicated before, medical applications represent only one market segment of the total potential biochip usage. Identical developments are also to be expected in the biotechnology, pharmaceutical, epidemiological, agricultural and environmental market segments. On their turn, biochips form only a small category of MEMS/NEMS devices.

Wireless (RF) technology, in combination with smart sensors/actuators and distributed computing, is gaining a lot of interest and offers huge potentials for both current and future environmental, machine and human monitoring and control. Section 3.6 in Part I presents details on wireless communication.

Finally, along with the growing number of MEMS and NEMS sensors and actuators is the increasing need for standardisations to support compatibility and exchangeability. This will be a time-consuming task, since it requires the full cooperation of all involved parties: application, electrical, software and telecom engineers.

15.3.4   Conclusion

It has been apparent for some time that the Moore's law is running out of steam due to the exponentially rising costs of IC design and manufacturing. In the meantime new technologies have been explored and developed. Silicon-based micro- and nanoelectronic technologies, micro- and nanoelectromechanical systems (*MEMS* and *NEMS*), microchemical systems (*MiCS*) or microbiological systems (*MiBS*), sensors and/or actuators are seen today as key technologies with a potential comparable to that of microelectronics. The combination of these new sensor/actuator technologies with microelectronic fabrication onto the same substrate will enable real system-on-a-chip solutions.

An increasing variety of materials used, together with greater complexity and new functionality of micro- and nano-devices, require new manufacturing processes, in which 3-D etching and stacking technologies will be of increasing importance.

All in all, *more than Moore* developments create huge opportunities and challenges, which require the involved semiconductor companies to build strong relationships with potential customers in these emerging markets and to expand their technology and design capabilities to beyond traditional chip development.

References[1]

1. "Worldwide Contract Manufacturing Review", <mfgmkt.com/wp-content/uploads/2015/10/July-2015.pdf>
2. Harry J.M. Veendrick, "Nanometer CMOS ICs, from Basics to ASICs", ISBN 978-3-319-47597-4, Springer 2017
3. "UltraScale Architecture and Product Data Sheet: Overview", https://www.xilinx.com/support/documentation/data_sheets/ds890-ultrascale-overview.pdf; February 15th, 2017
4. Shilpi Birla et al. "Analysis of the data stability and leakage power in the various SRAM cells topologies", International Journal of Engineering Science and Technology, Vol. 2(7), 2010, 2936–2944, <www.ijcst.info/docs/IJEST10-02-07-112.pdf>
5. Jeffery W. Butterbaugh et al., "Front End Processes", FUTURE FAB International I Issue 28, January 2009, pp. 75–81, <www.immersionlab.nl/pdf/FF28_Jan_09.pdf>
6. "A Better Computing Experience", https://www.sandisk.com/home/ssd, Sandisk, 2017
7. Empirical Evaluation of NAND Flash Memory Performance, <www.sigops.org/sosp/sosp09/papers/hotstorage_10_desnoyers.pdf>
8. "More on Future of Toshiba 3D NAND Flash Memory; BiCS will reach for 100 layers and beyond"; https://www.storagenewsletter.com/2016/08/11/more-on-future-of-toshiba-3d-nand-flash-memory/, August 11th, 2016
9. Kevin Gibb, "First Look at Samsungs 48L 3D V-NAND Flash", EE Times, April 6th, 2016
10. Martin Jared Barclay "Electrical Switching Properties of Ternary and Layered Chalcogenide Phase-Change memory Devices", <http://scholarworks.boisestate.edu/cgi/viewcontent.cgi?article=1066&context=td>
11. Rick Merritt, "3D XPoint Steps Into the Light", EE Times, Jan 14th, 2016
12. Chris Mellor, "Just ONE THOUSAND times BETTER than FLASH! Intel, Micron's amazing claim", July 28th 2015, The Register, http://www.theregister.co.uk/2015/07/28/intel micron 3d xpoint/
13. ISSCC conference web site: <http://isscc.org/> navigate to: Advance Program
14. Phillip Allen, "The Practice of Analog IC Design", <www.ewh.ieee.org/r6/scv/ssc/Allen_Practice_Analog.pdf>
15. Infineon Technologies, "EMC and System-ESD Design Guidelines for Board Layout", <https://www.infineon.com/dgdl/Infineon-ap2402633_EMC_Guidelines.pdf-AN-v03_05-EN.pdf?fileId=5546d46255dd933d0155e32392f1090e>, Edition February 22, 2016

---

[1]*Note:* Not every reader has access to the published articles of microelectronic conferences and magazines. A lot of effort has therefore been given to refer to those publications that are directly accessible through web pages on the Internet. However, these data may be volatile because some owners update and change the contents on their web pages, so that some of the references below may only be accessible during a short time after the print of this book. Finally a lot more information on the various subjects can be found by searching the Web with the right entry, which can be easily extracted from the corresponding subject. Good Luck!

© Springer International Publishing AG, part of Springer Nature 2019
H. Veendrick, *Bits on Chips*, https://doi.org/10.1007/978-3-319-76096-4

16. "Samsung to Develop Advanced NAND Flash Memory Along with Toshiba", <www.azonano. com/news.asp?NewsID=18686>
17. Joaquin Romo "DDR Memories Comparison and Overview", <www.freescale.com/webapp/ sps/site/overview.jsp?code=784_LPBB_DDR>
18. Dave Sroka, "Differences between USB and 1394", <www.allbusiness.com/technology/inter-net-technology/1106702-1.html>
19. Katalin Popovici, et al., "Virtual Platforms in System-on-Chip Design", <http://webadmin.dac. com/knowledgecenter/2010/documents/POPOVICI-VP-ABK-FINAL.pdf>
20. Sylvie Barak, "450mm wafers just a distraction say fab execs", <www.theinquirer.net/inquirer/ news/1433658/450mm-wafers-distraction-fab-execs>
21. MEMC <www.memc.com/index.php?view=Epitaxial-Deposition->
22. Masaharu Tachimori, "SIMOX Wafers", Nippon Steel Technical Report No. 73 April 1997, <www.nsc.co.jp/en/tech/report/pdf/7304.pdf>
23. George Celler, et al., "Smart Cut, A guide to the technology, the process, the products", <www. soitec.com/pdf/SmartCut_WP.pdf>
24. Narayana Murty Kodeti, "White Paper on Silicon On Insulator (SOI) Implementation", <www. soiconsortium.org/pdf/SOI_Implementation_WhitePaper_Infotech_v2.pdf>
25. Samuel Fung, "SOI process technology for the newest generation of high performance CPUs", <www.advancedsubstratenews.com/v9/articles/soi-in-action/from-the-foundry.php>
26. Copied with permission from ASML from the following website: <www.asml.com/asmldot-com/show.do?ctx=10448&rid=10081>
27. Joe Kwan, "Sign-off lithography simulation and multi-patterning must play well together", http://www.techdesignforums.com/practice/tag/multipatterning/, August 12, 2014
28. Ed Korczynski, "EUV Resists and Stochastic Processes", Semiconductor Manufacturing & Design Community, http://semimd.com/blog/tag/euv/, March 4, 2016
29. Peter Singer, "Nanoimprint-Lithography-A-Contender-for-32-nm", <www.ferret.com.au/n/ Nanoimprint-Lithography-A-Contender-for-32-nm-n676715>
30. Peter Clarke, "Report: Toshiba adopts imprint litho for NAND production", EElTimes (Analog), June 07, 2016
31. Intel White Paper, "Introducing the 45nmNext-Generation Intel® Core™ Microarchitecture", <www.intel.com/technology/architecture-silicon/intel64/45nm-core2_whitepaper.pdf>
32. Laura Peters, "Gaining Control over STI Processes", <www.jysong.idv.tw/cu/articles/cu0011. htm>
33. James Kawski, et al., "Ion implant: a new enabler for 32nm and 22nm devices", <www. electroiq.com/index/display/semiconductors-article-display/355567/articles/solid-state-tech-nology/volume-52/issue-3/features/cover-article/ion-implant-a-new-enabler-for-32nm-and-22nm-devices.html>
34. Wang Zhengfeng, "Chemical Mechanical Planarization", <http://maltiel-consulting.com/ CMP-Chemical-mechanical_planarization_maltiel_semiconductor.pdf>
35. Scott E. Thompson, "Uniaxial-Process-Induced Strained-Si: Extending the CMOS Roadmap", IEEE Transactions on Electron Devices, Vol.53, No.5, May 2006, <www.thompson.ece.ufl. edu/Fall2008/TED-%20extending%20roadmap-01624680.pdf>
36. Min-hwa Chi, "Challenges in Manufacturing FinFET at 20 nm node and beyond", http://www. rit.edu/kgcoe/eme/sites/default/_les/Minhwa%20Chi%20-%20abstract %20Challenges%20 in%20Manufacturing%20FinFET.pdf
37. Dylan McGrath, "Few Surprises as Intel, GF Detail Process Technologies", EElTimes, 12/7/2017
38. ITRS Roadmap Report, 2015 edition, https://www.semiconductors.org/main/2015_ international_technology_roadmap_for_semiconductors_itrs/
39. Jean-Pierre.Schoellkopf, "ATRS: an alternative roadmap for semiconductors, technology evo-lution and impacts on system architecture", <http://tima.imag.fr/conferences/async/Technical_ Program/Tuesday/Invited_talk_2/Async%202006%2014march06.pdf>

40. Trevor Mudge, "Power: A First-Class Architectural Design Constraint", <www.eecs.umich. edu/~panalyzer/pdfs/Power__A_First_Class_Design_Constraint.pdf>
41. Richard van Noorden, "The rechargeable revolution: A better battery", Nature, International Weekly Journal of Science, 05 March 2014
42. http://batteryuniversity.com, 2017
43. "Analog and mixed signal test", A recent history, Suraj Sindia, Springer 2018
44. 2017 International Mixed-Signal Testing Workshop (IMSTW)
45. "Test Yield Models", <www.siliconfareast.com/test-yield-models.htm>
46. Application note number 5: "Choosing a yield model with "11 Select a yield model or default", <www.icknowledge.com/our_products/Applications%20note%20number%205.pdf>
47. L. Peters, "DFN: Worlds Collide, Then Cooperate", Semiconductor International, June 2005, <www.synopsys.com/Tools/TCAD/CapsuleModule/semi_int_jun05.pdf>
48. "Focused ion beam (FIB)", Tutorials, <www.fibics.com/fib/tutorials/>
49. Taqi Mohiuddin, "FIB Circuit Edit Becomes Increasingly Valuable In Advanced Node Design", Jan 24, 2014
50. JEDEC, Joint Electron Devices Engineering Councils, <www.jedec.org>
51. Darvin Edwards, Hiep Nguyen, "Semiconductor and IC Package Thermal Metrics, Texas Instruments, Application Report, December 2003–Revised April 2016, <www.ti.com/lit/an/ spra953c/spra953c.pdf>
52. G.Q. Zhang, et al., "Mechanics of Microelectronics", Springer 2006, <www.springer.com>
53. Kevin J. Hess et al., "Reliability of Bond Over Active Pad Structures for 0.13-μm CMOS Technology", <www.freescale.com/files/technology_manufacturing/doc/ECTC_2003_ BOND_OVER_ACTIVE_KH.pdf>
54. Elpida Completes Development of Cu-TSV (Through Silicon Via) Multi-Layer 8-Gigabit DRAM", <www.elpida.com/en/news/2009/08-27.html>
55. Rao R. Tummala, "Moore's Law Meets Its Match", IEEE Spectrum, June 2006, pp. 38–43, <http://spectrum.ieee.org/computing/hardware/moores-law-meets-its-match/0>
56. Pär Persson Mattsson, "Why Haven't CPU Clock Speeds Increased in the Last Few Years?", Comsol Blog, November 13, 2014, https://www.comsol.com/blogs/havent-cpu-clock-speeds-increased-last-years/
57. Brian Bailey, "Do Chips Really Work The First Time?" Semiconductor Engineering, February 13th, 2014
58. M. LaPedus, "Costs cast ICs into Darwinian struggle", EE Times, March 30, 2007, http://www. eetimes.com/electronics-news/4070811/Costs-cast-ICs-into-Darwinian-struggle
59. "Semiconductor Wafer Mask Costs", Anysilicon, 2016, http://anysilicon.com/semiconductor-wafer-mask-costs/
60. Christian Piguet, "Microelectronics for Systems-on-Chips", Coursebook CSEM, Neuchatel, Switzerland, 2015-2016
61. Saul Hansell, "Counting Down to the End of Moore's Law", The New York Times, May 22, 2009, <http://bits.blogs.nytimes.com/2009/05/22/counting-down-to-the-end-of-moores-law/>
62. "Retina implant 'cures blindness'", <www.nursingtimes.net/nursing-practice/clinical-special-isms/public-health/retina-implant-cures-blindness/5023396.article>
63. Majid Eshaghi, "Automotive sensors overview", August 2017, https://community.nxp.com/ docs/DOC-335099
64. "All that is wrong with autonomous car sensors and how do we fix the problems", October 20, 2017, @ArcticREDTero, https://www.arcticred.io/single-post/2017/10/20/All-that-is-wrong-with-autonomous-car-sensors-and-how-do-we-fix-the-problems
65. "Top misconceptions of autonomous cars and self-driving vehicles", Alexander Hars, Inventivio GmbH, Inventivio Innovation Briefs, Issue 2016-09 (Version 1.3)
66. "Our Smart Road Future: Automated and Connected Weather Stations" https://fathym. com/2017/01/smart-road-future-automated-weather-station/, January 2017
67. Duke University, "DNA Lab on a Chip", <www.azonano.com/nanotechnology-videos.asp? cat=19>

68. National Human Genome Research Institute, "Lab on a chip", <www.genome.gov/pressDisplay.cfm?photoID=20017>
69. Erik C. Jensen, et al., "A digital microfluidic platform for the automation of quantitative biomolecular assays", Lab on a Chip, 2010, issue 6, pp. 685, <http://pubs.rsc.org/en/Journals/JournalIssues/LC>
70. Gregory Mone, "How it works: The Endoscope Camera in a Pill", March 2008, https://www.popsci.com/how-it-works/article/2008-03/how-it-works-endoscope-camera-pill
71. Jon Cooper, "Lab-on-a-Chip Technology, as a Remote Distributed Format for Disease Analysis" University of Glasgow, <http://ubimon.doc.ic.ac.uk/bsn/public/Jon_Cooper.pdf>
72. Handel Jones, "Why migration to 20nm bulk CMOS and 16/14nm FinFETs is not a best approach for Semiconductor Industry", White Paper, International Business Strategies, Inc., January 2014

Index

© Springer International Publishing AG, part of Springer Nature 2019
H. Veendrick, *Bits on Chips*, https://doi.org/10.1007/978-3-319-76096-4

Printed in the United States
By Bookmasters